Brion van Over, Ute Winter, Elizabeth Molina-Markham, Sunny
Communication in Vehicles

Also of interest

Intelligent Multimedia Data Analysis
Siddhartha Bhattacharyya, Indrajit Pan, Abhijit Das,
Shibakali Gupta, 2019
ISBN 978-3-11-055031-3, e-ISBN (PDF) 978-3-11-055207-2,
e-ISBN (EPUB) 978-3-11-055033-7

*Signal and Acoustic Modeling for Speech and Communication
Disorders*
Hemant A. Patil, Amy Neustein, Manisha Kulshreshtha, 2018
ISBN 978-1-61451-759-7, e-ISBN (PDF) 978-1-5015-0241-5,
e-ISBN (EPUB) 978-1-5015-0243-9

Understanding Security Issues
Scott Donaldson, Chris Williams, Stanley Siegel, 2018
ISBN 978-1-5015-1523-1, e-ISBN (PDF) 978-1-5015-0650-5,
e-ISBN (EPUB) 978-1-5015-0636-9

Lie Group Machine Learning
Fanzhang Li, Li Zhang, Zhao Zhang, 2018
ISBN 978-3-11-050068-4, e-ISBN (PDF) 978-3-11-049950-6,
e-ISBN (EPUB) 978-3-11-049807-3

Brion van Over, Ute Winter, Elizabeth Molina-Markham, Sunny Lie, Donal Carbaugh

Communication in Vehicles

Cultural Variability in Speech Systems

DE GRUYTER

Authors
Brion van Over
Manchester Community College
PO BOX 1046, Manchester CT 06045-1046, U.S.A.
bvanover@manchestercc.edu

Ute Winter
General Motors – Advanced Technical Center Israel
7 HaMada Street, 4673341 Herzeliya, Israel
ute.winter@gm.com

Elizabeth Molina-Markham
Northeastern University
University Scholars Program & Office of Undergraduate Research and Fellowships
411 Richards Hall, Boston MA 02115, U.S.A.
e.molinamarkham@northeastern.edu

Sunny Lie
California State Polytechnic University
Building 1-319E, 3801 West Temple Ave, Pomona CA 91768, U.S.A.
slie@cpp.edu

Donal Carbaugh
University of Massachusetts Amherst N308
Integrative Learning Center
650 N. Pleasant St., Amherst MA 01003-1100, U.S.A.
carbaugh@comm.umass.edu

ISBN 978-3-11-051891-7
e-ISBN (PDF) 978-3-11-051900-6
e-ISBN (EPUB) 978-3-11-051915-0

Library of Congress Control Number: 2019947838

Bibliographic information published by the Deutsche Nationalbibliothek
The Deutsche Nationalbibliothek lists this publication in the Deutsche
Nationalbibliografie; detailed bibliographic data are available on the Internet at http://
dnb.dnb.de.

© 2020 Walter de Gruyter GmbH, Berlin/Boston
Printing and binding: CPI books GmbH, Leck
Cover image: Metamorworks/iStock/Getty Images

www.degruyter.com

Acknowledgements

This book has involved a team effort in the best sense of that phrase. As authors, we have worked together to conduct the research presented here. The formation of our research team is discussed in detail in the first chapter. While we have been central to the research reported here, we have also benefitted from all kinds of support. At General Motors, in Michigan, our works were aided immeasurably by Timothy J. Grost, Laura Rosenbaun and Yael Shmueli; in China, we worked carefully with Peggy Wang. We were also helped in some translations by Xinmei Ge of Acton, Massachusetts and Libin Hang of Donghua University, Shanghai.

This research was supported by grants from General Motors to Donal Carbaugh at the University of Massachusetts. The authors are grateful for this support. Carbaugh is also grateful to the Department of Communication, the College of Social and Behavioral Sciences and the university for its continued support of his research.

Earlier versions of some of the works presented here have been published: Carbaugh, D., Winter, U., Molina-Markham, E., Van Over, B., Lie, S. and Grost, T. 2016. A Model for Investigating Cultural Dimensions of Communication in the Car. In: Theoretical Issues in Ergonomics Science 17(3):304-323; Carbaugh, D., Winter, U., van Over, B., Molina-Markham, E. and S. Lie. 2013. Cultural analyses of in-car communication. Journal of Applied Communication Research 41(2):195-201; Carbaugh, D., Molina-Markham, E., van Over, B., and U. Winter. 2012. Using communication research for cultural variability in human factor design. In: (N. Stanton, eds) Advances in human aspects of road and rail transportation. CRC Press. Boca Raton, (FL), pp. 176–185; Molina-Markham, E., van Over, B., Lie, S. and D. Carbaugh. 2015. "OK, talk to you later": Practices of ending and switching tasks in interactions with an in-car voice enabled interface. In: (T. Milburn, ed.) Communicating User Experience: Applying Local Strategies Research to Digital Media Design. Lexington Books. London. pp. 7-25; Molina-Markham, E., van Over, B., Lie, S. and D. Carbaugh. 2016. "You can do it baby": Cultural norms of directive sequences with an in-car speech system. Communication Quarterly 64(3):324-347; Wang, P., Winter, U., Grost, T. 2015. Cross cultural comparison of users' barge-in with the In-vehicle speech system. In: (A. Marcus, ed) Design, User Experience, and Usability: Interactive Experience Design. DUXU 2015. Lecture Notes in Computer Science, 9188. Springer; Winter, U., Tsimhoni, O., and T. Grost. 2011. Identifying cultural aspects in use of in-vehicle speech applications. Paper presented at the Afeka AVIOS Speech Processing conference, Tel Aviv, Israel; Winter, U., Shmueli, Y., and T. Grost. 2013. Interaction styles in use of automotive interfaces. In: Proceedings of the Afeka AVIOS 2013 Speech Processing Conference, Tel Aviv, Israel; Rosenbaun, L., Winter, U., and van Over, B. 2019. Voice Persona: Apologies in In-car Speech Technologies. In: (M. Scollo and T. Milburn, eds.) Engaging and Transforming Global Communication through Cultural Discourse Analysis. Fairleigh Dickinson University Press. Vancouver, B.C.

https://doi.org/10.1515/9783110519006-202

pp. 35-53; van Over, B., Molina-Markham, E., Lie, S., & Carbaugh, D. 2016. Managing interaction with an in-car infotainment system. In: (N. Shaked and U. Winter, eds.) Design of multimodal mobile interfaces. De Gruyter. pp. 145-168.

Contents

1 Cultural Analyses of In-car Communication

1.1 Introduction

When we sit with our laptop, phone, or television, we are involved in an interaction with communication technology. We have our own ways of thinking, our culturally based conceptions about that technology, about what it is, how it might be of use, and also, of course, about how we will indeed use it. This general arena of activity is called by some, the human–machine interface. This is of course a complex site of activity, and of study, as complexities of use lead to research reports with reports leading to the redesign of our technological devices or machines. This cycle of human use and research results in devices being the way that they are, and thereby sets the stage for end-users to use them in the complex and at times unexpected ways that we do.

One such device is being placed in the dashboard of cars or automotive vehicles generally. This sort of device allows drivers and passengers to adjust the temperature of the vehicle or the volume of speakers, to play a variety of radio stations, to play from personal music libraries, to make telephone calls, to navigate to destinations, and to conduct other such activities including the possibilities of messaging, conferencing, and other various forms of entertainment. Increasingly, multiple modes of interacting with such devices are being used. Earlier designs of these devices have relied mostly on push-button technologies. More often, now, a variety of modalities is integrated into a multimodal interface, such as touch screens or touch pads. When voice activation is used and when this is initially successful, for many tasks, users tend to like it.

The engineering of this technology for the dashboard of cars has been finely studied, as well as interactions between drivers and passengers in the car; however, the interaction between the people in the car and a dashboard device has been somewhat less studied, especially the cultural variety in its conception, use, and interpretation. For various reasons, among them the complexity of tasks and missing knowledge, a human–machine interface is developed for a large market, such as the US, and subsequently localized to a variety of other markets. For speech interface utterances of the machine, the responses to the user are often translated literally from one language to another, focusing on a generic dialog rather than cultural appropriateness, unconsciously violating cultural norms for communication. One observed example is the translation of the system English request "Please say your command" to the German "Bitte sagen Sie ein Kommando," which ignores the different meanings of Kommando in German* where the word is used solely for military purposes or for training pets and implies a boss–underling relationship and thus a notion of arbitrariness. General Motors (GM) has decided to invest into knowledge about cultural communication practices in important markets to offer a compelling human–machine interface for more pleasant user experiences. The practical difficulties GM investigators needed

https://doi.org/10.1515/9783110519006-001

to address, then, were cultural differences in not only uses but the recognition of languages including dialects, cultural differences in how errors were noticed then corrected, as well as cultural differences in the flow of in-car dialogue from task initiations to completions. What if new cars could adjust in-car devices to particular ways of speaking around the world? Investigators at GM Research and Development in Herzeliya, Israel, and in Warren, Michigan, have been pioneers in examining the human–machine interface and recently in noticing how cultural features like these were active at every stage but were not adequately being studied (Tsimhoni, Winter, & Grost, 2009). Dr. Ute Winter was quick to notice the need for basic research in this area and produced a call by GM for such research to address this variability. In it she wrote: "The goal of this research is to develop a framework of cultural dimensions and principles, which have influence on discourse and may lead to different perception of dialog success by conversation partners with different cultural background. This conceptual framework should enable GM to derive a method for empirical learning about culturally driven user expectations, decisions and behaviours, while interacting with speech applications in specific regions of the world." Dr. Winter sent this call to Carbaugh whose research, reported in Cultures in Conversation (2005) she had noticed as perhaps relevant to the call. To cut a long story short, Carbaugh, his team, and Winter with her team collaborated to produce a theory and methodology for doing such work, and subsequently conducted field studies using it in the United States and China (Carbaugh, Molina-Markham, van Over, & Winter, 2012). At GM, as a part of this project, the results of the field work are transformed into design considerations and recommendations for communication with future in-car infotainment systems using speech among other modalities (Winter, Tsimhoni, & Grost, 2011).

1.2 The Ethnography of Communication: Cultural Discourse Analysis

Derived from and indebted to the ethnography of communication (Hymes, 1972; Philipsen & Coutu, 2005), cultural discourse analysis is devoted to the description and interpretation of communication practice (Berry, 2009; Carbaugh, 2007; Scollo, 2011). A special focus has been the exploration of intercultural interactions, especially as this sort of variability in practice is active in specific cultural scenes, like, perhaps surprisingly, the automobile.

Given the need for attention to cultural variability in the human–machine interface, General Motors sought a perspective and methodology for examining such variability within in-car communication. Given the approach to its study, ethnographers of communication and cultural discourse analysts were well-suited to design and conduct such study.

The approach designed specifically for this project treated the automobile as a communication situation, with special attention to cultural sequencing of talk, uses of directives, opening and closing of tasks, and repair or corrective exchanges. In addition, special attention was given to the multimodal capacities of users as they used speech and touch. Similarly, our conceptualization was attentive to participant gaze as it moves from the road to the device and back again. The methodology for collecting data was designed to ensure as much comfort for participants as was possible. Each participant used their own vehicle which we equipped, in each case, with a tablet device running the in-car system with capacities for phone calling, radio-playing, access to a music library, and eventually navigation. The participant's car was equipped also with three cameras, one focused on the tablet device, one focused on the user's face, with the third offering a wide-angle view of the participant, the researcher in the front passenger's seat, the tablet, and the road ahead. All user interaction with the device and others in the car was thus recorded for purposes of our eventual analyses. This theoretical stance and methodology have produced extremely rich corpora of data from the United States and China.

Benefits of these studies went in multiple directions. General Motors sought an approach specially designed for the study of cultural features and dimensions of in-car communication that Carbaugh and Molina-Markham produced. A second stage of the project focused on the conduct of field studies in the US and in China. The research team, acknowledged and listed as co-authors of this book, then used the approach to conduct the field studies in each site. The benefits to the research team icluded support of this complex research project, collaboration with an interdisciplinary team which included linguists, engineers, human factor specialists, interaction designers, and others, access to GM's process of developing in-car technology, support for conducting fieldwork in multiple sites, and of course the opportunity to bring such work to completion through several collaborative efforts.

1.3 Criteria for Deciding to Participate

The process of deciding whether to assemble a team to respond to GM's call was not a simple matter. Several considerations animated that process. A primary consideration was whether the primary participants, the principal investigators, in this case Winter and Carbaugh, were "on the same page" with regard to the intellectual problems needing addressed and the general approach desirable for addressing them. In detailed discussions and correspondence, Carbaugh was assured Winter was knowledgeable about and supportive of the research that was needed. Winter, in turn, was confident the University of Massachusetts team could design and execute the type of research she wanted and needed to get done.

Another set of concerns related to the logistical support available for doing such work. After initial discussions, Winter asked Carbaugh what was needed, and Carbaugh responded with a tentative list of items, which Winter then adjusted, and so on. This process involved both principal players in producing short- and long- term planning for the project, which, in the long run, allowed the work to be done with adequate support.

A further set of concerns had to do with access to data. Carbaugh had worried that GM might put limits on the availability of the data collected and the construction of research reports. Winter was quick to convey that such limits were minor at the most and access would be assured throughout the entire project for analysis of data and research serving GM. Any concerns Carbaugh had were expressed by him immediately and were addressed directly by Winter in a timely way (and vice versa). So, the criteria used to decide whether to participate had to do with the synchronicity among participant researchers of the intellectual problems being addressed, the approach needed for addressing them, confidence by each party of the other, congeniality in relations among the key participants, adequate financial and logistical support for the project so designed including not only conceptual development but field studies, access to the data gathered, and few if any restrictions on the production of research reports.

From Carbaugh's view, from the start of the collaboration to the present day, the project provided an interesting set of problems to study, a unique set of communication practices to theorize about, and an opportunity to support graduate student researchers and colleagues, with Lie, Molina-Markham, and van Over falling into the former category, and Professor Libin Hang into the latter.

1.4 Difficulties and Challenges

At the same time, there were difficulties and challenges the project has faced. As readers of this book and this section in particular are undoubtedly aware, research projects must undergo review typically by an institutional review board (IRB). When a project puts study participants behind the wheel of an automobile with advanced infotainment technology to use, a number of questions are raised. An added factor is that the studies here are naturalistic or ethnographic, relying predominantly upon a qualitative research design, rather than the quantification of variables, and one can guess that this approach was not typical for such a review. Suffice to say here that the IRB review process was detailed, complicated, but eventually successful!

Throughout the research process, and in particular during the fieldwork phase, it was clear that we were operating under a set of tensions that needed careful balancing. One such tension existed between our ethnographic commitment to naturalistic study, and the needs of GM to acquire data and produce analyses that they knew could lead to actual revisions of the technology. For instance, as noted earlier, each vehicle was outfitted with a touch screen device that served as the visual and tactile interface to the "brain" of the in-car system. After installing this system in the participant's car, we provided them an opportunity to familiarize themselves with the system before going out on the road, and also during this time we collected data on how a new user of a system goes about learning what that system does and interacts with it.

The in-car system itself, provided by GM, posed a challenge on this undertaking. While the aim was to allow the user to formulate his intentions and desires from the infotainment system in his preferred natural language at any point of the interaction, and at the time of field research project, no in-car production system available on the market had these communication capabilities. This problem was solved by modifying an existing infotainment system, similar in design and manual-visual capabilities to the Cadillac CUE, and disabling the speech recognition capabilities in favour of an embedded Wizard, who would translate the user's spoken utterances into touch sequences in a specially designed Wizard interface. This solution guaranteed control over the performance of the system, while still most of its parts were a machine with the user communicating with that machine.

There was variation in the initial exploration and learning process while using this system. This meant that users had significant variance in their understanding of the system's abilities from the start. Some failed to discover that while doing one thing, listening to the radio for instance, they could initiate a speech command to do something else, like make a phone call. Such task-switching events are of interest to GM because they are scenarios in which misunderstanding is possible and in which preferences may exist for how the car accomplishes that task. Should the system mute the radio when a user asks to make a call, or simply dim the music? If a user never determines that this is possible and never decides to engage such a sequence, then these data will obviously not be collected. How then, do we allow users to use the system in whatever ways "naturally" make sense to them, which are data of great value, while also assuring that they have enough information about the system to try things they might want to try if only they knew they were possible?

Eventually, we decided that the best solution would be to break our observations of the drive into three phases. In the first phase, users explored with little to no input from researchers in a safe environment like a parking lot while the motor was idling. In the second phase, users attempted to employ what they discovered about the system's abilities in the parking lot, while out on the road, and sometimes discovered new abilities along the way. Around the halfway point of the ride, by prearrangement, we had the driver pull off the road. If by that time the user still had not discovered one of the system's primary functions, we asked prompting questions like, "Would you like it if the car were able to switch between radio and a phone call through a voice command?", "How might you ask the system to do something like that?", and "Do you want to give that a try?" In the third phase, then, users had the opportunity to drive back to the starting point, and, if so inclined, to try some ways of interacting with the system that perhaps they had not thought possible. These distinct phases allowed us to capture data of essentially different types and later analyze them as such, so as to exercise some minimal bracketing of our explicit influence. This tension played out in other arenas as well when questions arose about the extent to which we might manufacture problems to see how people would deal with them. Here the line between experiment and ethnography becomes problematically blurred and not crossing this line required vigilance and negotiation between members of the research team to assure all needs were ultimately met without compromising the integrity of the research, which all valued.

The human–machine interface, as represented in interactions with newly developing multimodal in-car systems, is a rich site of study for researchers interested in culturally distinctive communication practices. Automotive designers benefit as well from investigations into naturalistic usage of new designs. The work we presented in this chapter represents a collaboration between researchers and designers at various locations that led to the development of an approach for studying cultural features of in-car communication, data and reports from field studies in various sites, and design considerations and recommendations for future in-car systems. We discussed our participation criteria, as well as concerns and tensions that may arise when balancing commitments of naturalistic study with design development goals, emphasizing the importance of awareness, attentiveness, and cooperation in ensuring research integrity.

References

Berry, M. 2009. The social and cultural realization of diversity: An interview with Donal Carbaugh. Language and Intercultural Communication, 9:230–241.doi:10.1080/1470847090 3203058

Carbaugh, D. 2005. Cultures in conversation. Lawrence Erlbaum. (NJ)

Carbaugh, D. 2007. Cultural discourse analysis: Communication practices and intercultural encounters. Journal of Intercultural Communication Research 36:167–182. doi:10.1080/17475750701737090

Carbaugh, D., Molina-Markham, E., van Over, B., and U. Winter. 2012. Using communication research for cultural variability in human factor design. In: (N. Stanton, eds) Advances in human aspects of road and rail transportation. CRC Press. Boca Raton, (FL), pp. 176–185.

Hymes, D. 1972. Models for the interaction of language and social life. In: (J. J. Gumperz and D. Hymes, eds) Directions in sociolinguistics: The ethnography of communication. Blackwell. New York, pp.35–71.

Philipsen, G., and L. M. Coutu. 2005. The ethnography of speaking. In: (R. Sanders and K. L. Fitch, eds) Handbook of research on language and social interaction. Lawrence Erlbaum Associates. Mahwah, pp. 355–379.

Scollo, M. 2011. Cultural approaches to discourse analysis: A theoretical and methodological conversation with special focus on Donal Carbaugh's Cultural Discourse Theory. Journal of Multicultural Discourses, 6: 1–32. doi:10.1080/17447143.2010.536550

Tsimhoni, O., Winter, U., and Grost, T. 2009. Cultural considerations for the design of automotive speech applications. In: Proceedings of the 17th World Congress on Ergonomics IEA 2009, Beijing, China.

Winter, U., Tsimhoni, O., and T. Grost.2011. Identifying cultural aspects in use of in-vehicle speech applications. Paper presented at the Afeka AVIOS Speech Processing conference, Tel Aviv, Israel

2 A Model for Investigating Cultural Dimensions of Communication in the Car

2.1 Introduction

Extensive research has documented how the cultural nature of communication varies both within and across regions of the world (Carbaugh, 1990, 2005; Hymes, 1972; Philipsen, 2002). Speech enabled Human-Machine Interface (HMI) design needs to take into account this variation in user understandings of and preferences for different ways of communicating in different cultural contexts. Literature provides evidence for such cultural aspects of preferred and effective user interaction, though no theoretical framework exists to formally study cultural dimensions and their design implications in the automotive environment (Tsimhoni, Winter, and Grost 2009). In this paper, our primary purpose is to discuss a framework we developed for researchers and designers to study speech enabled HMI systems. This framework, or perspective for inquiry, is designed to discover the cultural nature of communication in contexts, while it also positions investigators to discover new cultural dimensions and principles which designers may not have considered. Although our framework can be adapted to a variety of communication contexts, we focus here on in-car communication of drivers with a speech enabled HMI.

We begin with an overview of key concepts. Then, we introduce how these offer synergy with User Centred HMI design principles. In the end, we demonstrate briefly the leverage our approach offers by discussing two results of our field studies concerning user interaction styles and how these resulted in re-designing HMI systems to accommodate user preferences.

We want to mention at the outset our concern about driver safety issues. A major concern of any interface design must be the maximal reduction of interference with the task of driving. The literature review and discussion of Barón and Green (2006) as well as Peissner, Doebler, and Metze (2011) show the advantages of a speech-enabled system on driving performance, but also mention its limitations. The quality of the system's design has a substantial influence on the potential for increasing drivers' safety. Therefore, the performance of the speech recognition system is crucial (Cooper, Ingebretsen, and Strayer 2014; Kun, Paek, and Medenica 2007). Equally important is the interface design itself for keeping the complexity of today's in-vehicle applications at a safe level (Zhang and Wei 2010) thereby avoiding driver confusion during communication (Cooper, Ingebretsen, and Strayer 2014; Maciej and Vollrath 2009).

Our studies have drivers using the most user-friendly, voice-activation possible. Furthermore, all have been conducted with drivers in their own cars, driving roads familiar to them which they select. This keeps drivers in their most familiar environ-

https://doi.org/10.1515/9783110519006-002

ment both within and outside the car. The framework that follows explores how driving and voice-activation work best together (and how they do not). While we do not intend to study cognitive workload directly, we aim for the design of a culturally appropriate experience. This will contribute to users' increased feelings of intuitiveness and naturalness in the use of such systems including turn taking, interaction style, error recognition and correction, among others. Our framework thus holds potential for reducing user confusion and unnecessary complexity. The eventual contribution of our studies to safer driving will result from our field studies and data analyses presented in the following.

2.1 The Theoretical Framework

In order to conduct field research, we have developed a theoretical framework for investigating the cultural dimensions and principles of communication which influence the different degrees of success people have in dialogue with a machine. This theoretical model includes a methodology for studying in the field culturally-driven user expectations, decisions and behaviours, as users interact with speech enabled in-car systems. The general approach, the theory and the methodology, is designed to be used, and has been used in specific regions, nations or communities, of the world (see for examples Milburn, 2015; Sprain and Boromisza-Habashi, 2013).

2.1.1 The Car as a Communication Situation

In order to understand cultural variations in the in-car Human-Machine Interface, we treat the car, interactions within it, and about it, as a "communication situation". In other words, we understand the car to be a situation understood through communication, and further, we understand communication situations to be at their base, culturally variable. The model derives from a long tradition in studies of cultural dimensions of communication from the seminal conceptualization of Hymes (1972), through Philipsen (1987, 2002), to Carbaugh (1988, 2007) among others. On these bases, we raise primary and fundamental research questions: What communication practices do people in fact do (and want to do) while in the car? Alternately, how do people talk about their car? How are these practices, and how is this talk culturally shaped and meaningful?

Our central construct, communication situation, includes several ingredients which we explore in response to these research questions: 1) the car is a place where people communicate with each other and with the car itself - in cultural ways; 2) the nature of that communication is done in ways which are distinctive to each speech community in particular; 3) those ways are structured through expressive norms, in other words, people want the interaction in, and with, the car to get done in some ways rather than in other ways; and 4) those distinctive ways, and those norms, activate users' preferences, which, if known, can help design this situation in ways that are pleasing to users. This is a sketch of the logic in our framework, which we ground with this construct of communication situation, and which we research specifically for human-machine communication with an in-car speech enabled HMI.

2.1.2 In-car Communication Events

Within the automotive communication situation, there are specific sequences of acts that can be understood as "communication events". In other words, each culturally situated, communication situation supports some communication events, rather than others. We understand a communication event to involve a sequence of communication acts, which – from the participants' view – has integrity as a sequence. For example, getting the car to play music by the Beatles can be understood as a communication event, as is getting the car to identify the closest-cheapest gas station, or similarly, getting directions to that gas station, or making a phone call. Each such task requires a sequence of communication acts, which participants understand to have some degree of cultural integrity – that is, it can (and should, they think) be initiated and completed in some ways rather than others.

From the view of our framework, we are using a nested conceptualization with communication situations including communication events which hold within them communication acts. We are developing the related point that communication acts in and about the car occur within events as "cultural sequences". In communication, parts of sequences are often identified by the language of participants as "greetings", "exchanging pleasantries", "thanks" and the like. Each sequence has some familiar flow or some sort of integrity to participants. Knowing the cultural sequences and the flow of a particular communication event, including how it is frustrated and/or corrected, as part of the larger communication situation, can add a deeper understanding of the communication people produce in the car. This is the logic in brief of our framework.

Within a communication event there are several types of "communication acts" we presume to be quite important to understand as humans interact in, and with a car. Some such acts are: opening, directing, addressing and referencing, closing, and also, repairing trouble. All occur within the natural and routine flow of interactions and involve cultural features such as specific forms, contents, and meanings.

Together with communication act sequences, our framework is designed to explore important extra-linguistic, multimodal cues. A few types of cues are gestural or tactile uses of the interface, facial expressions, and other nonverbal cues. These cues can include both prosodic features (intonation, stress, pitch, register) and paralinguistic aspects (tempo, pausing, and hesitation). These are purely indexical - or strictly tied to situations - because they do not necessarily have propositional content or context-free lexical meaning, but instead signal context-specific information. Different cultures have different ways of using or interpreting these "contextualization cues" (see Gumperz, 1982). If not properly understood as such, or if not considered as part of the communication situation, such cues can lead to misunderstanding.

A key concept coupled with multimodal cues is "conversational inference" (see Gumperz, 1982, 1992; Carbaugh, 2005). This concept brings into view how participants assign meaning to nonverbal cues within the on-going rhythm and flow of communication events.

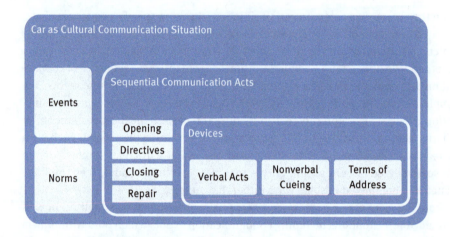

Fig. 1: Central Concepts in the Model

2.1.3 Design Dimensions for Speech Enabled HMI

Communication practices we are discussing here give a particular understanding to practical matters of HMI interaction design – which will become clearer in our discussion below. The design of an in-vehicle multimodal interface traditionally follows the paradigm of User-Centered Design (Norman and Draper, 1986; Nielsen, 1993; Vredenburg and Butler, 1996), which is widely considered the key to product usefulness and usability (Mao et al. 2005). Naturalistic observation, contextual inquiry, and other techniques (e.g., Holtzblatt, 2003; Holtzblatt, Wendell, and Wood, 2004; Dray and Siegel, 2007) are used to learn about user preferences in the relevant context, e.g. the driving situation and environment.

For now, consider that HMI interaction designers and researchers often develop a list of characteristics and parameters from daily work and experience, asking which of these are in question and which may vary cross-culturally. These practices or parameters are listed within our framework as dimensions of design in a way that they are observable, measurable and/or qualitatively identifiable in field data. The following summarizes several such dimensions which we have explored on the basis of our model. Each identifies potentially valuable information about cultural variability in speech enabled HMI. Our purpose in listing these is to demonstrate how our framework embraces and develops a wide range of such dimensions:

- **User satisfaction:** What do users think of the system? What parameters, behaviour, and instances lead to this opinion about the system?
- **User trust:** When and why do users trust or not trust the system?
- **Ease of use:** How well and intuitively can users understand the interface in a short time? What is easy? What is difficult?
- **User content:** How does the user find that the system deals with user content, such as contact lists, navigation data, local radio stations, etc.?
- **User interaction style:** How do users prefer to interact with the system? What is their personal communicative style? How do they expect the system to respond?
- **Multi-modal use:** Which modality do users prefer in what act or event? While driving on the road, are there things that a user feels he wants to do by touch, gestures, tactile or other modalities instead of speech? Do users feel that the visual feedback supports or is aligned with the voice interaction? What is irritating or misaligned rather than supportive?
- **Cooperative principles:** What makes the system cooperative or not cooperative in the practice of the user?
- **Turn taking:** Does a user know when it is his turn to interact with the system? In which situations does the user find it unexpected when the system responds or fails to respond to the user? How fast does a user take a turn? How fast should the system respond?
- **Grounding:** Do users think that the system and they have a shared understanding about the user requests and the system's capabilities to answer the requests? When do misunderstandings, conversational misalignments, and points of confusion occur?
- **Conflict resolution:** How do users find the system at dealing with repairs or misunderstandings?
- **Control handling:** When do users want the system to take initiative in communication? When does the user want to be in control?
- **Information distribution:** Do users want to tell the system their request in one utterance or rather let the system guide them and ask what it needs to know, even if the dialogue is longer?
- **Audible feedback:** How do users find the wording of the responses? How do users find the quality of the system's voice?
- **Context dependency:** How does user behaviour and expectation depend on the driving situation or environment?

Interaction dimensions of design such as these concern the verbal and/or non-verbal communication practices of the user while s/he is involved in a communication situation – or while executing a sequence of communication acts with the goal of completing a task. Each shows communicative practices that express an understanding, expectations and continuous learning of how the in-car interface works and communicates back. Therefore, while observing and analysing the user's communication practices in the car, we can learn about the user's expectations, seeing in an exacting way dimensions of interface design. In the end, as a result of these studied observations, we can derive design recommendations for interfaces in the in-car communication situation. Thus, we ground all design questions and areas in the framework of the larger communication situation, with special attention to specific communication events and acts. With this focus, we are attentive to local norms or preferences which we theorize as follows.

2.1.4 Norms in Communication Situations and Events

As we have already noted, communication situations and events have been defined as the activities, or aspects of activities, that are directly governed by rules or norms for the use of speech. The concept of communication event draws attention to ways participants go about doing each interaction sequence and emphasizes that each can be done in a proper way according to participants preferred conduct within the car.

Norms are "statements about conduct which are granted some degree of legitimacy by participants" (Carbaugh, 2007, p. 178). They are messages about "correctness" that may be stated explicitly by participants or they may be more implicit as a dimension in the structure of participants' practices. In our view, a norm includes the essential element of an "ought" or a "should." In this sense, communication norms tap into a moral domain of practical action, practices which people believe should be conducted in some ways, rather than in others. This way of understanding communication norms, then, is not simply a description of a regular or normal routine (i.e., the thing people routinely do), but of a normative practice (i.e., what people believe should or should not be done).

Communication norms, as practical actions, can of course vary in their force (see Philipsen, Coutu, and Covarrubias, 2005; cf. Jackson, 1965, 1975). As we adapt Jackson's early ideas, the concepts of intensity and of crystallization help distinguish the variety of dimensions in and types of communication norms.

- **Intensity:** How strongly do people feel about the normative practice? On a Likert scale from 1-7, a bi-polar distribution with half at 6-7 strongly liking it, and half at 1-2 strongly disliking it may be possible. The intensity, then, regarding this norm is strong, with respondents registering the strength of their feeling at the extreme ends of the scale. Of course, if most respondents were in the 3-5 range, we discover that people report not feeling very strongly about this norm, or it is rather weak in its intensity.

- **Crystallization:** Through such analyses, we can know further if there is or is not general agreement with regard to this practice of communication. When there is a bi-polar distribution in the results – as when some favour it while others do not - the norm would have, as discussed above, a high degree of intensity since people feel relatively strongly about it, yet a low degree of crystallization since people do not agree on how it should be done.

Again as we adapt Jackson's earlier ideas, there are three types of communication norms which derive from these dimensions of intensity and crystallization:

1. **Conflict Potential:** These are norms with high intensity and low crystallization.
2. **Normative Power:** These are norms with high intensity and high crystallization.
3. **Vacuous Consensus:** These are norms with low intensity and high crystallization.

Knowing what are the topical matters of concern, through which norms, with what degree of importance, and the like, we are better equipped to design a communication environment which people feel is more their own, closer to the expectations and desires they hold. We refine our understanding of these situations by examining particular communication events including the norms or rules active within them.

The key objectives then, of the framework, are several: to discover specifically how in-car communication gets done; to describe human-machine interfaces generally, as communication situations which are culturally distinct and therefore cross-culturally diverse; to interpret participants' meanings of communication in these situations; to study comparatively the nature of communication in these situations; and to provide guidance in the design of communication in these situations.

We now discuss our methodology, generally. Our purpose here is to cast our treatment of the methodology as unattached to a particular field site thereby demonstrating its utility in any field site around the globe. We do note the framework and methodology have been fruitfully used and developed in the United States and in China (see Milburn, 2015; Molina-Markham et. al., 2016; Sprain and Boromisza-Habashi, 2013). Our treatment of the methodology is in two sections, the first focused on overall field design and data collection, the second on three phases of qualitative data analysis.

2.2 Research Design

2.2.1 Data Collection in the Field

Since human-machine interaction depends to a large extent on the communication situation, its communication acts, events and norms, we consequently need to create an environment in which users are able to act as natural as possible if we want to observe their practices, in this case, in the car environment. Any field study design must, therefore, be guided by the principle of "ethnographic or naturalistic inquiry", observing users in their natural car environment, driving where each wants, acting as each would like, and interacting with the car and its HMI in their preferred ways. Our general methodology involves a sequential research design in four general phases, with each phase involving a specific set of activities from the researchers' activities prior to entering the field, to activities completed after leaving the field.

2.2.1.1 Pre-fieldwork Activity

This phase of a project involves several activities, which are preliminary to doing the fieldwork itself. It is crucial to be as knowledgeable about a field site as possible. If there is a literature available about the site, then it can and should be consulted. What is the history of the area, its people, its economy, driving tendencies, occupations, and so on? It is helpful to have some local knowledge about where one is going and who are the participants in the field study. A second set of activities involves preliminary planning about the fieldwork itself. According to the special focus envisioned, a preliminary conceptual map can and should be formulated. Focusing on specific concerns establishes a theoretical position from which to observe communication in (and about) the car. This equips one for study and reflection while in the field, enables a systematic approach to one's observations, as well as provides ways of designing interviews (and eventually cars). We emphasize that this conceptualization – the framework and interview protocol – is not a closed design, to go unrevised after one is in the field, but an open, yet structured design to be reflected upon and revised as needed based upon subsequent field activities.

2.2.1.2 Fieldwork Activity

This phase of activity involves periods of observation, in this case of communication in the car itself. Observations such as these inevitably lead to subsequent questions that researchers want to ask participants. The accumulation of a descriptive record about observational and interview events creates a large corpus of data, which then needs to be analysed. The analysis involves the distinct modes of investigation discussed in the following section.

2.2.1.3 Post-fieldwork Activity (A)

Conduct detailed descriptive analyses first; then, conduct detailed interpretive analyses. The activities conducted after leaving the field involve deeper phases of analysis. These analyses lead, often, to additional questions about the dialogic in-car dynamics observed, or heard about while in the field. This phase of the research can lead, when possible, back to the field for more detailed observations and analyses. It can also lead to further studies, which are conducted under modified conditions, for example in a controlled experiment focusing on predefined parameters for observation, or in a simulated driving environment.

2.2.1.4 Post-fieldwork Activity (B)

Conduct comparative analyses, and critical assessments; write empirical field research reports; offer interface design and policy recommendations based upon findings. These final phases of a field project are crucial as each creates a sharper view of the cultural dimensions of communication getting done in, and about the car. These are often better understood through comparative analysis, for instance, by conducting the field studies in different cultures. Note, and we emphasize, that the phases we present here are sequential in their design, but cyclical in their possibilities. A phase of fieldwork can result in revising one's earlier conceptualizations; post-fieldwork activities can lead one back to the field focused on other observations, with different questions; writing a report can lead to its presentation, with that presentational event being an occasion for subsequent observing and interviewing; and so on. The general research design is linear but its typical implementation is cyclical (see Carbaugh and Hastings, 1992).

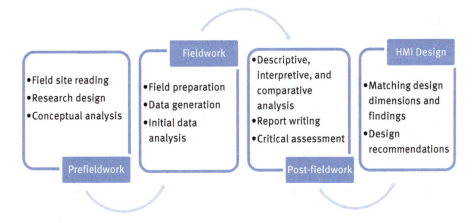

Fig. 2: A Brief Portrayal of the Field Methodology

2.2.2 Field Data Analyses: Descriptive, Interpretive, Comparative

In our discussion above, we have discussed one key purpose of our framework, collecting data in the field. Here, we address three - descriptive, interpretive, and comparative - ways those data can be subsequently analysed.

The above framework focuses on the car as a communication situation, and within it, specific events and acts which are culturally shaped. But how are these to be analysed? Here we provide a set of components for such descriptive analyses. For such descriptive analyses, there are 8 basic components we use to analyze a communication situation, event or act (see Hymes, 1972; Carbaugh, 2012). We will use the concept, communication practice, here, as an umbrella construct, which includes the concepts above, such as communication situation, communication event, communication act, and all of the dimensions we discuss of the HMI design. Following the central objectives of the framework we discussed above, the following components are used initially to describe the communication that is happening, and then subsequently to interpret the meanings of that communication to participants. The components also provide a basic general set of elements for comparing communication practices across communities, nations, or scenes.

For an initial **descriptive analysis**, the components function as a series of questions to be asked about communication practice thereby discovering how communication is done within and about the car. We summarize this descriptive use of the components as follows:
1. **Setting:** In what physical environment is the communication taking place?
2. **Participants:** Who is involved in the communication practice?
3. **Ends:** This component has two parts: What are the participant's goals of the practice (e.g., to send an email)? What are the outcomes of the practice (e.g., the email was sent, or the effort to do so was unsuccessful, or the user got irritated at the car)?
4. **Act/Sequences:** What specific communication acts got done, and in what sequence?
5. **Key:** What was the emotional pitch, or tone of the communication (e.g., perfunctory, serious, frustrated)?
6. **Instrumentalities:** What multiple mode(s) or cues were used in this communication (e.g., voice, gesture, pressing or touching a button)?
7. **Norms** (see last section): What were the norms – stated and/or implied – for this interaction?
8. **Genre:** Is there a generic form to this communication practice which participants use, and if so, what is it?

These 8 components provide a basic investigative framework for analysts to systematically describe any communication practice, such as communication in, and about the car.

Our general methodology is designed in distinct phases. A second phase of analysis focuses on **interpretive inquiry**. During this phase, the above framework (along with an additional set of analytic procedures) will again be used, but this time to interpret the participants' meanings of the communication practice described above. In other words, the same concepts can be used as follows:

1. **Scene:** What do participants say is "going on" in this setting, through this communication practice?
2. **Participant identities:** What culturally significant roles, identities, or relationships are active in this practice, from the participants' view?
3. **Ends:** What are the goals and outcomes sought by participants in this practice, in their own terms?
4. **Act/Sequences:** Is there a regular routine, and perhaps an ideal routine, or act/sequence for doing this? What is it? How "relaxed" or strictly formulaic is that sequence?
5. **Key:** How do participants interpret the feeling of this practice?
6. **Instruments:** What are the necessary, and/or preferred modes or channels for doing this activity?
7. **Norms:** What norms do participants employ for interpreting this activity?
8. **Genre:** Is there a generic cultural form active here, and can it be identified by participants?

In this second step, as in the descriptive analysis, the HMI design questions will be consulted to extract HMI-relevant insights from the data such as User Interaction Styles, and the preferences associated with those preferences, which we discuss later in the article.

The 8 components in relation to the HMI design dimensions are being used here, then, in two ways; first, to describe the activities in the car mainly from an analyst's point-of-view; then to investigate, and unveil the participants' meanings of those communication practices, in that situation, through their sense of the events, or through their ways of speaking. Consequently, the results will lead to design recommendations for the in-car speech enabled HMI.

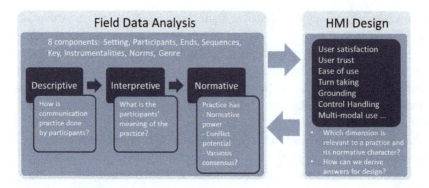

Fig. 3: Analysis Method of Field Data

A following, third phase, of comparative analyses respond to the questions: to what degree is this practice the same, here and there, and to what extent is it different? If one were to study direction-giving in Massachusetts and in Shanghai, or speech prompting of entertainment, we would expect some similarities and some differences in this practice in the two locations. Through such study, we get a better idea about what indeed is culturally distinctive in one set of practices, as opposed to those in another; we also get a better sense of what is similar across such practices. Through such comparative study, and based upon descriptive and interpretive analyses, we are able to determine what parts in the speech enabled HMI need a different design, and what parts can display the same behaviour in all examined cultures. While it is obvious that due to different languages the system prompts need to be transferred from one culture to the other, there may be less evident considerations related to dialogue flow, task directives, addressing, interaction style, communication tone, user trust, and others, which may have influence on the prompt design for each culture. For example, in our initial comparative studies, we have discovered less complex dialogue flow in directives (i.e. fewer communication acts or shorter sequences) in our Chinese data than in our US data.

2.3 Experimental Set Up for Field Research

2.3.1 An Embedded Wizard of Oz Interface for the Car

For our empirical studies, we replaced a user's in-car infotainment system with an embedded Wizard of Oz multimodal interface, providing the user with four typical domains for infotainment: phone dialling, radio tuning, music selection, and navigation. Graphic samples of the design employed in these contexts are provided in Appendix A. Passonneau et al. (2011) gives an overview on the use of embedded wizardry for dialogue systems including a brief discussion of system accuracy when a human wizard replaces part of the recognition process. Figure 4 shows our system mounted on the air vent in the car centre console and a view of a driver.

Fig. 4: User Perspective of Embedded Wizard of Oz HMI

The application allows the user two modes of interaction. The user can choose to conduct all tasks via touch screen in the centre console, when field study regulations permit. This kind of interaction will be handled by the application without intervention from the Wizard. The user may also choose to initiate a task event by speech after touching a speech icon on the screen. In this case, the task of the Wizard is to replace the Speech Recognizer and Natural Language Understanding module, transforming the user intention into a touch sequence on his Wizard screen, including the choice of confidence levels that a machine would produce.

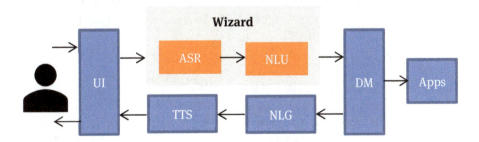

Fig. 5: Illustration of Embedded Wizardry into Dialogue System

Using such an embedded Wizard system encourages the user to behave most naturally and in the user's preferred way, while the front-end and dialogue management of the system is a machine and perceived as such by the user.

2.3.2 Multimodal Design Principles

The embedded Wizard system allows us to explore design principles, which cannot yet be implemented to this degree in speech applications due to technological limitations. Jurafsky and Martin (2009) or Jokinen and McTear (2010) give an overview of the status of speech technologies and dialogue systems.

As for spoken dialogue sequences, the Wizard system supports full Natural Language Understanding, flexible information distribution within a dialogue sequence, mixed-initiative dialogues, context dependency, and multimodal support. Consequently, the Wizard of Oz interface is highly cooperative. The user can take a turn at almost any point during a dialogue, he can interrupt the system, change his request, express himself freely, etc. Moreover, the Wizard can learn about user preferences and context during the study and use such knowledge to adapt the system's understanding to the user.

As for manual-visual interactions with the touch screen, the user interface relies on several design principles. Visual cues, such as colour, icons, location, and spacing/grouping elements, are used with the aim of improving the level of feedback to the user. According to the principle of cross-modal priming, visual prompts echo spoken prompts, but the visual display is also "stand alone". Another design principle is the use of shaping text. Once the user presses the speech button, example sentences appear to shape the spoken interaction and provide information of how to use the system. Finally, vertical and horizontal sets of tabs serve to provide visual context for speech communication.

The Wizard of Oz system is a fully symmetric multimodal infotainment system that integrates both visual-manual capabilities, as well as speech input and output throughout a flexible interactive experience, and allows switching between modalities at any point of an interaction sequence. We discuss next one way the above theory and methodology was applied.

2.3.3 One Brief Example of a Field Data Collection Process

One of our field projects was conducted in western Massachusetts in an area that contained primarily rural roads, as well as some urban and suburban driving. The roads in this area do not typically experience a high volume of traffic. Twenty-six participants were selected so as to represent a wide variety of characteristics, 12 male and 14 female drivers. Nineteen of the participants were originally from the northeastern

United States. Participants ranged in age from 26 to 64 years old and had between 1 and 48 years of driving experience. The average income of participants was around $65,000 (US dollars). Participants spent on average around 9 hours driving per week. Nineteen of the participants were smartphone users, and 7 were not. Participants were recruited through flyers posted in public areas or through snowball sampling procedures as a result of direct contact with researchers. The overall session with each participant lasted about 120-150 minutes. (Our other field sites were in urban Shanghai and Beijing, China, with the same research approach being used in both.)

Participants brought their own car to the study, as well as their personal content to the Wizard of Oz interface, such as the address book of their mobile phone or their preferred music collection. As two team members installed the Wizard system and observation equipment for visual and audible documentation (which records the participant's face, the tablet screen, the dashboard of the car and the view out of the front windshield), the participant was introduced to the experiment and given the consent form. The field research team typically included 2 researchers in the car: the "experimenter" in the front passenger seat and one researcher, the "Wizard", operating the Wizard of Oz interface in the back seat. A third researcher helped with the equipment installation.

Once in the car, the participant was introduced to the basic capabilities of the infotainment system and was then invited to explore interacting with the system through visual-manual means and specifically through speech, while experimenting with using the system for a variety of tasks. Then the participant drove around a large, open, off-road area until comfortable with the tested interface. Afterwards the participant performed a 60-75 minute drive of their choosing. During the drive, the participant was encouraged to use the Wizard of Oz interface in his preferred way with the exception of not using touch, other than touching the speech button or ending an application.

After completing the drive, the driver was asked a series of questions according to an interview guide relating to the dimensions of the dialogue experience, including turn taking, task completion, error correction, quality of audible feedback, multimodal use, general ease of use, and user satisfaction, among others. All questions were open questions (Winter et al., 2011).

Finally, the researchers met together in a debriefing session to reflect upon the specifics of the driving session, the interview session, particular observations made during each, and to identify useful focal concerns for further attention and analysis.

2.4 Summary

How a person chooses to interact with the various forms of technology that surround him or her is deeply connected to the cultural context in which those interactions take place. In this paper, we have explicated a theoretical framework and its associated methodology for conducting empirical field studies that explore the cultural dimensions of individuals' communication with speech enabled interfaces in their car. This model includes key concepts from the ethnography of communication and cultural discourse theory including communication situation, event, and rules or norms. These are used in conjunction with several design dimensions to produce HMI-relevant findings regarding HMI design dimensions. In this tradition, the experimental set up and analyses are designed to reveal qualitative as well as quantitative results. We briefly demonstrated one productive finding of our research concerning one HMI dimension, styles of user interaction.

One finding from our field studies includes these two different styles--one with a focus on efficiency in task completion and a second that emphasizes elaborate interactivity and a personal relationship with the system--that are preferred by participants to varying degrees in a northeast region of the United States. We add at this point that our field data collection in China, and its comparative analyses, revealed the same two general interaction styles, with differences between the two mainly being in the language itself including the dialogue flow relative to its sequential structuring. This reveals differences in directive formulation, error corrections, and event structure generally. Using these styles as guides, among other culturally distinct features (see this research teams' other published field studies), researchers can develop speech enabled systems that more closely align with user expectations and preferences, here demonstrated briefly to illustrate interaction flows and prompt design. In terms of future work, the field data we have collected is invaluable, deeply rich and detailed enough to facilitate the development of the exact phrasing of prompts, the detailed specification of the voice persona for both interaction styles, the flow and rhythm of the system, as well as development of the other design dimensions discussed above.

References

Barón, A. and P. Green. 2006. Safety and usability of speech interfaces for in-vehicle tasks while driving: A brief literature review. Technical Report UMTRI-2006-5. University of Michigan Transp. Res. Inst.

Carbaugh, D. 1988. Talking American: Cultural discourses on DONAHUE. Ablex. (NJ)

Carbaugh, D. 1990. Cultural communication in intercultural contact. Lawrence Erlbaum. (NJ)

Carbaugh, D. and S. O. Hastings. 1992. A role for communication theory in ethnography and cultural analysis. Communication Theory 2(2):156-165.

Carbaugh, D. 2005. Cultures in conversation. Lawrence Erlbaum. (NJ)

Carbaugh, D. 2007. Cultural discourse analysis: Communication practices and intercultural encounters. Journal of Intercultural Communication Research 36:167–182. doi:10.1080/17475750701737090

Carbaugh, D. 2012. A communication theory of culture. In: (A. Kurylo, ed) Inter/Cultural Communication: Representation and Construction of Culture, Sage. Thousand Oaks, pp. 69-87.

Cooper, J. M., Ingebretsen, H., and D. L. Strayer. 2014. Mental workload of common voice-based vehicle interactions across six different vehicle systems. AAA Foundation for Traffic Safety. Washington (DC)

Dray, S. and D. Siegel. 2007. Understanding users in context: An in-depth introduction to fieldwork for user centered design. In: (C. Baranauskas, P. Palanque, J. Abascal, S.D.J. Barbosa, eds) Human-Computer Interaction – INTERACT. Lecture Notes in Computer Science. Springer. Berlin Heidelberg, 4663: pp. 712-713

Gumperz, J. J. 1982. Discourse strategies. Cambridge University Press. Cambridge.

Gumperz, J. J. 1992. Contextualization and understanding. In: (A. Duranti and C. Goodwin, eds) Rethinking context: Language as an interactive phenomenon. Cambridge University Press. Cambridge, pp. 229-252.

Holtzblatt, K. 2003. Contextual design. In: (J. Jacko and A. Sears, eds) The Human-computer interaction handbook: Fundamentals, evolving technologies and emerging applications. Lawrence Erlbaum, Mahwah, pp. 941-963.

Holtzblatt, K., Wendell, J.B., and S. Wood. 2004. Rapid contextual design: A how-to guide to key techniques for user-centered design. Morgan Kaufmann. San Francisco (CA)

Hymes, D. 1972. Models for the interaction of language and social life. In: (J. J. Gumperz and D. Hymes, eds) Directions in sociolinguistics: The ethnography of communication. Blackwell. New York, pp.35–71.

Jackson, J. M. 1965. Structural characteristics of norms. In: (I. D. Steiner and M. Fishbein, eds) Current studies in social psychology. Holt, Rinehart and Winston. New York, pp. 301-309.

Jackson, J. M. 1975. Normative power and conflict potential. Sociological Methods and Research 4: 237-263.

Jokinen, K. and M. McTear. 2009. Spoken dialogue systems. In: (G. Hirst. Morgan and Claypool, eds) Synthesis Lectures on Human Language Technologies, 2(1):1-151.

Jurafsky, D. and J. H. Martin. 2009. Speech and language processing: An introduction to natural language processing, speech recognition, and computational linguistics. 2nd ed. Upper Saddle Prentice-Hall. River (NJ)

Kun, A., Paek, T. and Z. Medenica. 2007. The effect of speech interface accuracy on driving performance. Proceedings of the 8th Annual Conference of the International Speech Communication Association (Interspeech '07), Antwerp, Belgium.

Maciej, J. and M. Vollrath. 2009. Comparison of manual vs. speech based interaction with in-vehicle

Mao, J., Vredenburg, K., Smith, P., and T. Carey. 2005. The state of user-centered design practice. Communications of the ACM. 48 (3):105-109.

Milburn, T. (ed.). 2015. Communicating user experience. Lexington Books. London.

Norman, D. and S. Draper. 1986. User centered system design; New perspectives on Human-Computer Interaction. Lawrence Erlbaum. Mahwah (NJ)

Nielsen, J. 1995. Usability engineering. Morgan Kaufmann Publishers Inc. San Francisco (CA)

Passonneau, R.J., Epstein, S.L., Ligorio, T., and J. Gordon. 2011. Embedded wizardry. In: Proceedings of the SIGDIAL 2011 Conference. pp. 248-258.

Peissner, M., Doebler, V., and F. Metze. 2011. Can voice interaction help reduce the level of distraction and prevent accidents? In: Meta-Study Driver Distraction Voice Interaction. Carnegie Mellon University, White Paper. Pittsburgh (PA)

Philipsen, G. 1987. The prospect for cultural communication. In: (D.L. Kincaid, ed) Communication theory: Eastern and western perspectives. Academic Press. San Diego, pp. 245-254.

Philipsen, G. 2002. Cultural communication. In: (W. Gudykunst and B. Mody, eds) Handbook of international and intercultural communication. Sage. London and New Delhi, pp. 51-67.

Philipsen, G., Coutu, L.M., and P. Covarrubias. 2005. Speech codes theory: Restatement, revisions, and response to criticisms. In: (W. B. Gudykunst, ed) Theorizing about intercultural communication Sage. Thousand Oaks, pp. 55-68.

Sprain, L., and D. Boromisza-Habashi. 2013. The ethnographer of communication at the table: Building cultural competence, designing strategic action. Journal of Applied Communication Research, 41: 181-187.

Tsimhoni, O., Winter, U., and Grost, T. 2009. Cultural considerations for the design of automotive speech applications. In: Proceedings of the 17th World Congress on Ergonomics IEA 2009, Beijing, China.

Vredenburg, K., and M. Butler. 1996. Current practice and future directions in user-centered design. In: Proceedings of Usability Professionals' Association Fifth Annual Conference, Copper Mountain (CO)

Winter, U., Tsimhoni, O., and T. Grost.2011. Identifying cultural aspects in use of in-vehicle speech applications. Paper presented at the Afeka AVIOS Speech Processing conference, Tel Aviv, Israel.

Zhang, H. and L. Ng Wei. 2010. Speech recognition interface design for in-vehicle system. In: Proceedings of the 2nd International Conference on Automotive User Interfaces and Interactive Vehicular Applications (AutomotiveUI '10). ACM, New York (NY)

3 "You can do it baby": Non-Task Talk with an In-Car Speech Enabled System

In this chapter, we focus on directive sequences, and the phenomenon of participants from the northeastern United States using non-task talk. The analysis of these sequences reveals a norm that one ought to engage in non-task talk with the system. We suggest that this norm is grounded in a user premise that the system's interactional status involves the ability to speak. We find that this norm lacks crystallization among participants, and we formulate a competing norm that helps to account for this. The second norm reveals an underlying belief that the system's status as a machine is the basis for how it should be treated. We compare these findings with an analysis of data from mainland China.

3.1 Introduction

In this chapter, we perform cultural discourse analysis to examine directive sequences in which participants interact with an in-car speech enabled interface. In human-human interaction, the act of giving a directive (for example, telling another person to act in a certain way) has implications for the relationship between individuals, and thus for issues concerning considerations of face and politeness (Goffman, 1959; Lakoff, 1973; Grice, 1975; Brown & Levinson, 1978; Searle, 1990; Shahrokhi & Bidabadi, 2013). Giving a directive has the potential to impinge upon the sense of self of the person to whom the directive is given, so how one chooses to formulate a directive reveals a great deal about the speaker's view of the addressee (Blum-Kulka, 1997). Research on interactions between people and computers has suggested that people will often interact with a system similarly to how they would interact with a person, even though they may recognize that their interactional partner is non-human (Nass & Brave, 2005; Turkle, 2011). In our analysis, we find that two competing cultural norms are active in the interaction when drivers engage with a speech enabled interface in their car—one norm that one ought to engage in non-task talk with the system and a competing norm that non-task talk is unnecessary—and we suggest that the norms we identify activate cultural premises of personhood that constitute the car as an interactional partner in distinctive ways. In other words, how participants go about telling their in-car system what to do speaks about how they view that system and its potential for interaction—that is, whether they view the non-human system as similar to a human interactant or not.

These competing cultural norms in communication have likely developed as a result of new speech enabled technologies that force participants to consider different types of interactional partners, such as computers. Whereas previously, it was often possible to engage with these types of technology through tactile means, such

https://doi.org/10.1515/9783110519006-003

as pressing a key or touching a screen, interactions through voice are becoming more prevalent in daily life, and individuals must now consider what type of interactional partner a computer actually is or indeed what they want it to be. These findings have significant implications for the development of new culturally-adaptive speech systems and for increased understanding of how people orient to technology more generally, such as in the form of their smartphones or tablets.

3.2 Research on Design, Human-Machine Interaction, and In-Car Communication

Our examination of cultural norms in communication brings together several research areas, including research on design, on human-machine interaction, and on in-car communication between humans. Jokinen and McTear (2009) provide an overview of the design and development of spoken dialogue systems, dividing these systems into two main types: "task-oriented systems" which "involve the use of dialogues to accomplish a task," and "non-task-oriented systems" which "engage in conversational interaction, but without necessarily being involved in a task that needs to be accomplished" (p. 1). According to Jurafsky and Martin (2009), the goal of the field of speech and language processing is "to get computers to perform useful tasks involving human language, tasks like enabling human-machine communication, improving human-human communication, or simply doing useful processing of text or speech" (p. 1). Models and theories for research on speech and language processing draw from research in computer science, mathematics, electrical engineering, linguistics, and psychology, among others, in order to design systems that can use a knowledge of language in order to accomplish speech recognition, understanding, and synthesis. Speech recognition, understanding and synthesis require knowledge of phonetics and phonology, morphology, syntax, semantics, pragmatics, and discourse. Models and algorithms are used to resolve ambiguities in language usage so that computers can determine the meaning of utterances by determining such elements as parts of speech, the sense in which words are used, and sentence types.

In recent years, the design and development of systems for interaction with humans has moved away from an engineering-driven approach and toward user-centered design (UCD), or "a multidisciplinary design approach based on the active involvement of users to improve the understanding of user and task requirements, and the iteration of design and evaluation" (Mao et al., 2005, p. 105). Designers view UCD as a way to make systems more useful and usable by taking into account how people use technology in their larger environmental context (Holtzblatt & Beyer, 2014). For example, previous research has emphasized the general importance of cultural context in design. Based on case studies in India and South Africa, Mäkäräinen, Tiitola, and Konkka (2001) propose the need to consider that cultural factors in design should go beyond language issues. Drawing on a case study in Namibia on cultural biases concerning the concept of usability, Winschiers and Fendler (2007) argue that design methods result in locally inappropriate evaluations of usability. They suggest that researchers should not rely on their own assumptions about usability, but instead should actively and explicitly confirm through empirical studies the contextual meaning of these criteria in different places.

Reviewing research on human-machine interaction reveals that in many situations people treat machines similarly to human interactants (Reeves and Nass, 1996; Friedman, 1997; Jurafsky & Martin, 2009; Nass & Yen, 2010). Jurafsky and Martin (2009) write "It is now clear that regardless of what people believe or know about the inner workings of computers, they talk about them and interact with them as social entities" (p. 8). According to Nass and Brave (2005), numerous experiments show that "the human brain rarely makes distinctions between speaking to a machine—even those machines with very poor speech understanding and low-quality speech production—and speaking to a person" (p. 4). In her research on human relationships with sociable robots, Turkle (2011) describes the current state of human interaction with computers as the "robotic moment" (p. 9). With this phrase, she describes users' willingness or "readiness" to treat robots as valid partners with whom they can have relationships (p. 9). Turkle (2011) argues that people are not necessarily deceived into thinking technology is alive, but they are willing to "fill in the blanks" and interact with robots as if they could form a relationship with them because robots "perform understanding" (p. 24-25). Turkle's (2011) concept of a robotic moment emphasizes that how a robot interacts is more important in shaping how some individuals will respond to the robot, than is the fact that the robot is not a person. As a consequence, within the field of speech and language processing, some researchers focus on the design of "conversational agents," or "artificial entities that communicate conversationally" (Jurafsky & Martin, 2009, p. 8).

Prior research on in-car communication has primarily focused on human-human interaction. For example, researchers have examined fine-grained sequencing of human interactions within the car (Laurier, 2005; Laurier, Brown, & Lorimer, 2007; Laurier et al., 2008; Haddington, 2010), the interactional management of distractions in the car (Haddington & Keisanen, 2009; Koppel et al., 2011), and how speaking within the car is tied to social roles (Laurier, 2011; Laurier, Brown, & Lorimer, 2012). However, research has only just begun to explore interaction between humans and speech interfaces in the car as a site of cultural variation (Tsimhoni, Winter, & Grost, 2009; Winter, Tsimhoni, & Grost, 2011; Carbaugh et al., 2012; Winter, Shmueli, and Grost, 2013). This chapter builds on this past work in order to explore cultural norms that are active in these situations (Carbaugh et al., 2013). One of the cultural norms—that of using non-task talk in interaction with the in-car system—suggests a case in which human-machine interaction parallels human-human interaction. We note, however, that this norm of interaction seems to lack support among some users, perhaps because competing norms exist that are informed by differing cultural premises about the proper relationship between machines and humans.

3.3 Approach

As noted in chapter 2, cultural norms are statements, which, implicitly or explicitly, are granted legitimacy in a speech situation or community (Carbaugh, 2007). Norms vary in strength and influence (Hall, 1988/1989, 2005; Jackson, 1975). We build on Jackson's (1975) concepts of intensity and crystallization to help distinguish the variety of norms. Intensity refers to how strongly interactants feel about a norm—for example, do participants feel very strongly that a particular norm should or should not be followed, or are they relatively indifferent about it? Crystallization, on the other hand, refers to the general agreement among participants about a particular norm. Thus, for example, there may be norms about which participants feel strongly, but, if there is a split between those who strongly support and those who strongly do not support the norm, we would say that the norm lacks crystallization because there is not general agreement about it. We formulate norms in a prototypical four-part formula. This formula addresses: 1) where the norm takes place (the setting or context); 2) who the person is engaging in the norm (or what role they are taking on); 3) what the strength or force is of this norm; and 4) what action is being accomplished (Carbaugh, 2007). Formulating norms in this way makes explicit participants' expectations for how an interaction within this type of situation should unfold and provides a basis for comparison with other communication situations, such as those involving in-car systems in other cultures.

Communication events are made up of communication acts, such as the giving of a directive, which is a type of communication act designed to get the hearer—in this case the car—to do some action (Ervin-Tripp, 1976; Ervin-Tripp, Guo, & Lampert, 1990; Searle, 1990; Fitch, 1994; Blum-Kulka, 1997; Goodwin, 2006). We should note that in our analysis we are using the concept communication act as it is used in the ethnography of communication—see the exchange among John Searle, Michelle Rosaldo, and Dell Hymes (in Carbaugh, 1990). An analysis of directives as such can reveal users' considerations of face and politeness (Goffman, 1967; Lakoff, 1973; Grice, 1975; Brown & Levinson, 1978; Shahrokhi & Bidabadi, 2013). Goffman (1967) proposed that personal conduct during an interaction is "the combined effect of the rule of self-respect and the rule of considerateness" (p. 11). In other words, a concern for one's own face and for the face of one's interlocutors guides a person's actions. Building on this idea, Brown and Levinson (1978) used the notion of "politeness" to draw attention to the systematic reasons users will deviate from a basic principle of efficiency. Similarly, Grice's (1975) Cooperative Principle includes four maxims (quantity, quality, relevance, and manner) that provide a basis for understanding times when speakers may seem to violate one of these maxims (such as quantity, by giving more information than is necessary) in order to create meaning through inference. According to Lakoff (1973), speakers may in these cases of violation be following a politeness rule. Lakoff (1973) proposed "rules of pragmatic competence," including that one ought to "be clear" and "be polite" (p. 296). Her first rule of clarity drew on Grice's maxims. Lakoff's (1973) second politeness rule includes three sub-rules that one ought to not impose, to give options, and to make the addressee feel good (p. 298). Blum-Kulka (1997) explained the connection between an analysis of directives and issues of face and politeness, noting that "All types of social control acts impinge on the recipient's freedom of action and constitute a threat to face; therefore, politeness becomes a major consideration in the choice of mode of performance" (p. 142). Thus, the concepts of face and politeness form a basis for our formulation of norms of communication with an in-car system. We link the idea of a communication act (giving a directive) with the larger interactional sequence of which it is a part (a directive sequence) and wider cultural norms users assume for facework and politeness, for proper enactment in and with the car.

3.4 Data Analysis

Our analysis seeks to explore the characteristics of the communication situation when a driver interacts with his or her in-car speech enabled system and to identify communication norms and cultural premises that are active in order 1) to understand such dynamics, and 2) to develop suggestions for the design of future technology. We begin with an overview of the directive sequences that we analyzed. We then explore the phenomenon we found of non-task talk that occurred during directive sequences. We focus on dynamics of turn-taking and functions of non-task talk, as well as participants' own reflections. These examinations are the basis for formulating norms of interacting with an in-car speech system—in other words, statements about conduct that resonate with participants' understandings of how this interaction should take place—which will explicate participant expectations, provide a basis for comparison with other cultures, and enable the application of this knowledge to the design of future systems.

3.4.1 Directives

Participants would initiate a speech interaction by touching the microphone button on the tablet screen. Touching the button would cause a chime or "ding" to sound, and the microphone button would become illuminated with a green light to indicate that the system was ready to receive verbal input. Participants would then speak to the system in order to accomplish one of the possible tasks of listening to the radio, listening to their music, or making a phone call. We analyze these participant utterances as directives. As Goodwin (2006) explains, directives are "utterances designed to get someone else to do something" (p. 515). Searle (1990) argued that speech acts varied by three primary dimensions: by the expressed psychological state of the speaker, by the point or purpose of the act, and by the fit of the words to the world. In the case of a directive, the act expresses something that the speaker wants to happen, and the point or purpose of the act is to get the hearer-car to do something. The speaker is attempting to use words to create a new state of affairs.

Researchers have examined directives in terms of directive sequences (Fitch, 1994; Goodwin, 2006), which we found in our data. Below is an example of one such directive sequence. It occurred at the beginning of Participant 23's on-road drive, approximately two and a half minutes after he had pulled onto the highway.

Instance 1: Participant 23 - 21:47
P = Participant, S = System
1　P:　((touches microphone button and system dings)) My music.
2　S:　What kind of music would you like to hear?
3　P:　Artists.
4　S:　What artist would you like?
5　P:　LMFAO.
6　S:　One moment please. (2.0) Playing lmfao or select an album.
7　　　((music plays))

Directive sequences varied in length and complexity. In reviewing our data, we noted that at times participants produced talk during directive sequences that did not seem directly relevant to accomplishing a system task. This led us to question the form and function of this type of non-task talk in interaction with the in-car system.

3.4.2 Non-task talk in directive sequences

In this section, we present the results of our analysis of 79 instances of non-task talk in directive sequences. These instances were produced by 15 participants. We use the term task to refer to the system's action of making a phone call, playing a radio station, or playing a song. By non-task talk, we mean participant talk within the directive sequences that was not directly connected to telling the system what to do. In other words, this talk was not part of the minimum requisite information for the system to accomplish one of its tasks. There is an important difference here between what another person could understand versus what a speech enabled system can understand. Furthermore, there is a distinction to be drawn in the analysis of human-computer interaction between speech that is formulated with an expectation of response and that which is not. We apply this distinction here to task talk, which we suggest is formulated as hearable by, and relevant to, the system's ability to accomplish the task at hand, and non-task talk, which, while it may occupy a conversational position that in human-human interaction would necessarily inform the next speaker's turn, does not do so here as the system is not assumed by the speaker to be "listening" or equipped to understand the utterance.

We want to emphasize that the 79 instances of non-task talk analysed here all addressed the system, rather than the two researchers in the car. Many of these instances included terms of address that selected the in-car system, rather than the researchers, as recipients. We identified instances when participants addressed their utterances to the system by finding when participants used the pronoun "you" following a system utterance, for example when drivers thanked the system or encouraged the system. We also understood the system to be the intended recipient of the utterance when a participant said please to the system immediately after a directive or when a participant answered a question posed by the system. However, in future research, it would also be interesting to consider instances of non-task talk that appeared less obviously addressed to the system, as well as to consider the role that researchers played as audience members or intended overhearers of non-task talk (Goffman, 1981; Gordon, 2013).

Following the completion of data collection, recordings were consulted for instances of non-task talk. Some participants produced only one instance, while others produced many. Directive sequences that contained non-task talk were transcribed and analysed in terms of sequencing and functions. The mid- and post-driving session interviews were consulted for instances when the researcher in the front seat asked participants to reflect back on what they had said when interacting with the system. These reflections informed analysis of the transcribed instances, and together, both transcribed instances and interview data were the basis for formulating cultural norms and premises of communication.

Line 5 in Instance B below is an example of non-task talk that occurred during Participant 7's off-road testing session. In this instance, Participant 7 expressed appreciation for the system having returned to the "home" screen as she had requested.

Instance 2: Participant 7 - 25:35

1 P: Can you go back to the original? ((turns to the researcher)) Can I
2   give it a name?
3   ((system changes to Home screen))
4 P: Oh!
5 S: OK.
6 P: That was good! Good for you, System! Oh, OK, I'm going to call you
7   Denise.

Participant 7's statement in line 6 was addressed to the system, which she referred to as "you," "System," and "Denise." It was an instance of non-task talk in that it was not directly related to giving the system information so that it could accomplish a task (because the system had already accomplished its task). Instead, this particular instance of non-task talk appeared to serve relational functions of congratulations.

Non-task talk also sometimes occurred throughout a directive sequence and not just at the end. The instance below occurred after the mid-session interview, about thirty seconds after the participant had started to drive again. In it, Participant 18 responded to the system's request for him to "Hold on" with "You bet." The system did not require information at that point as it undertook the task of playing a jazz station, until it asked the question "What music station do you want to hear?" at which point Participant 18 provided the needed information (task talk) of "Real jazz" in line 5. Similarly to Participant 7, Participant 18 expressed appreciation in line 7 when the system accomplished the task of playing 67 XM Real Jazz, even though at that point the task was accomplished and no more information was required.

Instance 3: Participant 18 - 51:35
1   P: ((touches microphone button and system dings)) Uh, jazz.
2   S: Hold on.
3   P: You bet.
4   S: What music station do you want to hear?
5   P: Uh, what was it called? Real jazz?
6   S: Tuning radio to 67 XM real jazz.
7   P: Thank you.

Given that the non-task talk was not required for the system to accomplish its tasks, the question arises of what role this talk serves in the interaction and how this connects to participants' understandings of the communication situation. Here we draw further on cultural discourse analysis to analyse this non-task talk in terms of some of its key characteristics, including turn-taking and functions, and to explore how participants themselves accounted for their talk. We then use this analysis as a basis for formulating two competing norms of communication that appear to be active when participants in this context talk with and about their in-car speech enabled system.

3.4.2.1 Turn-taking
One notable characteristic of the non-task talk demonstrated in the data is that participants seem to use this non-task talk to coordinate turn-taking in interaction with the system. For example, in some cases, participants appeared to employ non-task

talk to indicate that the system was not starting its turn at the appropriate time—either it was speaking too quickly or not quickly enough. For example, in lines 4 and 13 of Instance 4 below, which took place after the mid-session interview about three minutes after the participant had started to drive again, Participant 9 implied that the system had taken too long to respond.

Instance 4: Participant 9 - 50:13
```
1   P:  ((touches microphone button and system dings)) Let's listen to
2       WNNZ.
3       (4.0)
4   P:  Come on, what's taking you so long.
5       ((Researchers indicate that the system needs a restart and refresh
6       the system.))
7   P:  ((touches button and system dings)) <Let's listen to WNNZ.>
8   S:  Please wait.
9   P:  OK
10      (10.0)
11  S:  Just a second.
12      (8.0)
13  P:  That's a second.
14  S:  Please review your station. It may not be valid.
15  P:  Why not?
16      (5.0)
17  P:  <WNNZ> 91.7=
18  S:  =Pardon?
19  P:  91.7
20  S:  Tuning radio to . . .
```

In line 2, Participant 9 questioned why the system was taking a long time. This comment did not provide the system with needed information for completing its task, and the system did not respond to Participant 9's statement. However, at this point the researchers did restart the system, suggesting that they may have been responding to Participant 9's utterance as possibly expressing concern or frustration. In line 13, Participant 9 told the system, "That's a second" after approximately eight seconds of silence, implying that the system had taken too long. Again, this did not include information necessary to the task at hand that the system could respond to.

Another example of non-task talk being used to coordinate turn-taking occurred in the following instance, which took place after the mid-session interview about a minute after the participant had started to drive again. In this instance in line 4, Participant 14 told the system, "I'm thinking, shut up," indicating that the system had not allowed enough time between turns.

Instance 5

Participant 14 - 1:06:39

1 P: ((touches microphone button and system dings))
2   (8.0)
3 S: Pardon?
4 → P: I'm thinking, shut up.
5 S: Could you repeat that please?
6 P: Um, play Rage Against the Machine- actually play Tool.
7 S: Just a second. (7.0) I am sorry. I am unable to find the music item
8   that you are looking for.
9 P: ((Participant presses the End button.))

While Participant 14's utterance in line 4 could be understood as a directive in that he was telling the system what to do, this utterance did not provide information that the system needed to accomplish its task, and there did not appear to be an expectation of reciprocity on the part of Participant 14 that the system would respond to this utterance. The participant did not initiate a repair sequence when the system asked him "Could you repeat that please?" after he had told it to "shut up," but instead, continued with the interaction and asked the system to play music by a particular artist. This suggests that the user did not have an expectation of response, or system comprehension in the formulation of this utterance, and was, therefore, not oriented to the accomplishment of the task.

Researchers have found that delays in responding to requests in interactions between people often precede rejections, whereas grantings are not delayed in this way, because of a preference principle for avoiding or minimizing disagreements, disconfirmations, and rejections (Pomerantz & Heritage, 2013). Thus, participants who responded to system delays with non-task talk may have been interpreting the system's delay as an inability to accomplish the directed task—in other words, a rejection of the directive. This interpretation was apparent in the instance below when Participant 13 said, "No you don't like that one?" in line 3, following approximately nine seconds of silence. This instance occurred about twenty minutes after the participant had started the on-road driving session and approximately ten minutes prior to the mid-session interview. Participant 13's question about liking her directive did not provide information that could help the system accomplish its task. In line 10, she called the system a "goofball," which was again not something that the system required in order to act, though it could be hearable as an evaluation of the system's performance in task accomplishment. In lines 13 and 17, Participant 13 encouraged the system, and in line 18, she expressed appreciation for the system's action, similar to Participants 7 and 18 above. Again, this final statement was not related to telling the system what to do since the task had already been accomplished.

Instance 6

Participant 13 - 39:53

1 →	P:	((touches microphone button and system dings)) Play XFM.
2		(9.0)
3		No you don't like that one?
4	S:	Please wait.
5		Ah
6	P:	(9.0)
7	S:	Please review your station. It may not be valid.
8	P:	((laughs)) Ok mm. (5.0) Play my music.
9	S:	Please let me know what you want.
10 →	P:	Ok. You are a goofball. (4.0) I'm going to [call you-]
11	S:	[Wait a] moment. What
12		kind of music would you like to hear?
13 →	P:	My music. You can do it. Come on.
14	S:	What song do you want to hear?
15	P:	Hmm. Play Creep.
16	S:	Hold on.
17 →	P:	You can do it baby.
18		Playing Creep. ((music plays))
19 →	P:	There you go.

In this case, Participant 13 indicated in line 3 that she believed the system was unable to accomplish the task she had given it of playing the radio by interpreting the silence as the system "not liking" her directive. Her non-task talk could have been an attempt to determine if she should reformulate her directive so that it could be accomplished. However, while a human interactant may have been able to interpret her question about "liking" a directive as an attempt to obtain more information about why there was a delay, the system could not, and her non-task talk did not provide additional information that would have allowed the system to accomplish its task.

3.4.2.2 Functions

In addition to coordinating turn-taking, there is evidence in the data of participants employing non-task talk to accomplish several functions. For example, participants used the non-task talk to encourage or support the system. This function is evident in Participant 13's statement, "You can do it baby" in Instance 6, which was said after she had asked the system to play music, and it had responded with "Hold on." In addition, the non-task talk in these instances often consisted of an evaluation of the system's actions. Frequently this evaluation was positive, as when participants acknowledged that something had been done correctly—for example, "That was good! Good for you, System" said by Participant 7 in Instance 2 at the end of the sequence after the system had accomplished its task. The function of evaluating the system's actions seems to align with the participants' understanding that they were participating in the driving session in order to test a prototype, as was explained to them before the driving session.

Evidence that participants oriented to the system as something that they had been asked to test can also be found in the explanations that participants gave to the friends and family members whom they called during the session. Several participants asked the people they called about the sound quality of the call. They stated that they were driving with researchers in their car and testing a new system. For example, when Participant 5's mother answered the phone and commented that Participant 5 had appeared on the caller ID as "anonymous," Participant 5 responded, "Yes, I'm testing out an in-car system thing. So, that would be why." During his driving session, Participant 17 left a message on the voicemail of one of the people whom he called, saying, "Hi [name]. I'm calling you from my car on a phone- on a computer. So I'm curious to know what the sound quality is like." In response to this explanation, call recipients would then provide feedback about how the driver sounded on the phone—Participant 8's mother told her that it was difficult to hear her, and said, "So you might give that feedback to your researchers." Thus, drivers viewed their role to be that of testing the in-car system. Their non-task talk can be understood as further constructing their relationship with the system as one who has been asked to evaluate it.

In this context, participants generally did not address the system angrily or yell at the system. This may have been largely due to the fact that the researchers were also present in the car, and participants did not feel comfortable speaking negatively in front of the researchers, as this might imply a negative evaluation of the researchers' system. Thus, while participants were instructed not to directly address researchers during the driving session, it is possible that some system-addressed non-task talk might have been indirect messages designed for researchers. A similar dynamic has been identified by Tannen (2004) and Gordon (2013) who examined ways that interactants used speech directed at pets and technology to accomplish certain identities, such as "compliant study participants" (Gordon, 2013, p. 306). Just as participants were constructing their relationship with the system as one who has been asked to evaluate it, they were likely, in addition, constructing an identity of evaluator for the researchers' benefit by showing themselves to be cooperative and helpful evaluators. We focus here on participants' understandings of their relationship with the system, but future analyses could explore how participants use speech enabled technology to negotiate relationships with other people, such as researchers or others in their car, and to create identities in interaction.

In two of the instances included above, participants did speak critically to the system. Participant 9 in Instance 4 asked the system, "what's taking you so long," and Participant 14 in Instance 5 told the system to "shut up." Interestingly, both of these instances took place after the mid-session interview when participants were encouraged by the researchers to talk to the system as they would do if they had been alone. These instances may indicate that participants would likely speak to a system in their car in both a positive and a negative manner, if they were not being observed by researchers. Although Participant 13's statements to the system were generally positive during the session, she claimed during the mid-session interview that if she were to have this system in her car on a daily basis, she would speak to it both positively and negatively. When the interviewer observed that during the driving session, Participant 13 had apologized to the system, Participant 13 explained,

> Well I would also yell at it. I mean I would treat it like a person. I'm generally nice and polite, but then, if it was being smart-ass, I might yell back at it. But I wouldn't, you know, I wouldn't get mad. Not really.

As this quote suggests, we must be careful not to overestimate the emotional force of critical remarks directed at the system, since at least for some participants while they might yell at it, they "wouldn't get mad. Not really." This comment is further suggestive of the potentially playful nature of some in-car interactions with the system.

3.4.3 Participant reflections on interacting with the system

In the tradition of CuDA, we examined not only communication occurring with the in-car system, but also communication about the system—such as Participant 13's comment about speaking negatively above—for evidence of cultural norms and premises being used in interaction, in particular conceptualizations of the system as interactional partner. The mid- and post-session interviews provided useful sources of communication about the system. During these interviews, several participants expressed that they felt the need to produce some non-task talk to the system. Some participants described producing non-task talk to the system as "tempting" or "natural." For example, Participant 9 said during the interview that he thought "it's natural to sort of, converse with it, you know, as I would a person." Participants observed that they talked to the system without really meaning to or thinking about it. When the interviewer commented during the interview at the end of the session that Participant 22 had said "all right" in response to the system's saying "Wait a moment," the participant noted that she often speaks to systems that produce speech, such as her GPS, in this way. She explained,

> I always do that to these. I almost like jokingly, like "Oh OK, thank you." Like whenever it says, "You have arrived." I'm like "All right! Thank you!" . . . I mean it's so anthropomorphized that like, I don't know. I'm used to it, I guess.

When the interviewer noted during the mid-session interview that Participant 15 had said "thank you" to the system and asked if that is the way he would interact with the system, Participant 15 commented,

> When you're talking to somebody and they do something, you say thank you. . . . I kind of thought to myself after I said it, I was like, "Jesus, dude. It's not like she has a real name." . . . I guess it just popped out. I was like "Thank you." It was, it just seemed like it was a natural reaction to asking somebody to do something for me.

Participant 15's mentioning of the characteristic of having a "real name" reinforces Participant 22's comment that the system has some human-like qualities. It is also significant that Participant 15 refers to the system with the pronoun "she," rather than "it," as this pronoun would also appear to indicate an orientation to the system as a female person, rather than as a computer. In this way, some participants found it natural to interact with the system using non-task talk, and they appeared to view the system to a certain extent as if "she" were a female person.

The key terms "nice" or "polite" were drawn on by participants as a way of describing and understanding their behavior when asked about it during interviews. During the mid-session interview, when the interviewer commented to Participant 20 that he had said "please" to the system and asked if it would be his preference to talk in this way with the system when using it, Participant 20 reflected, "Yeah, I mean, why not? It's helping you out. You might as well be nice to it." Similarly, when the interviewer noted that Participant 13 had apologized to the system, Participant 13 claimed, "Well, it's technology. That doesn't mean you don't have to be polite." At the end of the session, during the final interview, the interviewer observed to Participant 18 that he had inserted some phrases into the interaction with the system, such as "oh, great, thanks" or "oh, sure." The interviewer asked if Participant 18 inserted these types of comments for fun, and Participant 18 observed,

> No, I do it subconsciously. I'm just trying to be polite to the system. . . . You got to understand from a social science standpoint, I think we are a very impolite culture, and I'm trying to change that so, it sort of creeps in.

Thus, one way that participants had of describing their use of non-task talk was as being "nice" or "polite." Furthermore, some participants even admitted that they would feel uncomfortable not talking to the system in this way. At the beginning of the mid-session interview, the interviewer asked Participant 7 what her experience had been like so far interacting with the system. Participant 7 said that she had never interacted with this type of a system before in which she had to "give orders," and she did not really like the feeling that she was giving orders, or, as she also called it, "barking things at it." In other words, engaging in non-task talk with the system was, for Participant 7, a way to avoid feeling that she was giving the system orders. These interactional dynamics exhibit a participant stance of being a good, polite conversational partner—one both self-reflective and considerate of the other/ system.

There were participants, however, who did not share Participant 7's discomfort. A key idea that emerged in the data when those who did not produce non-task talk were asked about their interactions was that the system was just a "machine." For example, when asked during the interview at the end of the session about how she relates to her car and if she would prefer her car to speak to her as another person would, Participant 5 said that she thinks of her car as a "tool," and she does not see it as something that she would relate to as she would another person. She said,

> I think I see it more as a tool. I do not name my car, and you know, that type of thing. . . . I think of it still, still a functional tool, that, you know, I'm using for directions, or to call people, or radio. . . it-it's a machine, technology.

Similarly, during the mid-session interview when the interviewer asked Participant 16 about a complaint that he had because he felt the system was interrupting him, the participant observed that he felt that these interruptions emphasized for him the feeling that he was interacting with a computer and not another person. He went on to note, however, that this was "fine" because "I don't need her to be my friend." He explained,

> I don't need this to be human. I just need it to recognize what I tell it to do and do it. I don't need any of the, "Ok, see you later." You know, I don't take any like comfort from it having a couple of pre-programmed personable phrases. . . . I'd rather it just was straight and to the point.

Thus, although for some participants, it seemed difficult to not engage in what they perceived to be a "nice" manner with the system by using non-task talk, others recognized it as a "machine" and were not concerned with engaging in this type of talk. This distinction appears to be linked to a view of the system as either having some person-like qualities or as being non-human.

3.5 Findings: Competing cultural norms active during driving sessions

Based on the above analysis, we suggest that a key finding of this research is the presence of at least two competing cultural norms in participants' communication with and about the in-car voice activated system. We also propose that these two norms are based on cultural premises about personhood—or how participants understand being a person in this context.

One cultural norm that we found to be active in these types of interactional sequences could be formulated as:

1.   *when interacting with another, or a system participant (i.e., the in-car system), if this is done properly, one should engage in some non-task talk with the in-car system.*

In other words, the user can be understood to act so as to maintain the "face" of the in-car system by offering encouragement, such as "You can do it baby," acknowledging when something has been done correctly, or calling the system by a personal name, such as "Denise." This encouragement, acknowledgement, and naming could be understood as instances of what Brown and Levinson (1978) would call positive politeness strategies in that they indicate some approval of the system or its actions and emphasize intimacy with the system. In talking in this way, participants flout Grice's (1975) maxims by including more in their utterances than is required at that point in the interaction.

Participant 20 noted during the mid-session interview that the system, "talks to you a little bit, and, you know, takes on some humanish, you know, characteristics when she's talking." For Participant 20, being spoken to in a "humanish" way seems to merit a human response regardless of the material embodiment of that speech. Thus, we suggest that the above norm is based on a cultural premise of personhood that we formulated as:

2.   *social interaction is based on an agent's (i.e., the system's) way of communicating rather than its status as a human.*

This premise underlies drivers' non-task talk to the system, and also their reference to the system with the female pronoun, "she." Nass and Brave (2005) observe that users will tend to assign gender to a voice, even if the voice is clearly mechanically produced, responding to synthetic voices "as if they reflected the biological and sociocultural realities of sex and gender" (p. 15). In this way, for these participants, treating the system as a female person is connected to its speaking in a female human way, regardless of whether it is a person or not.

Although some participants treated the system similar to a human interactant in spite of its status of "non-human," our data indicate that even these participants did still recognize that "she" is not human. For example, some participants spoke negatively toward the system, although Participant 13 observed that even if she were to yell at the system she would not "really" be mad. This would seem to indicate that even though some participants are interacting with the system "as if it were human" and appear to demonstrate concern for the system's face through routine uses of politeness, the absence of expectation of response from the system to those forms of interaction free the user from serious concern over potentially face-threatening actions. This enables participants to "yell at the system," a risk that human interactants would be unlikely to take given the possible social consequences.

Participants' reflections on their interactions with the system during the sessions indicate a lack of crystallization, or agreement among participants, of a norm of engaging in some non-task talk with the in-car system. While participants, such as Participant 7, expressed feeling this norm strongly, to the point of discomfort if they did not engage the system with some non-task talk, many participants used relatively little to no non-task talk with the system, and when asked about their view of talking with the system, noted that it was just a "machine" or "tool." Given the novelty of this type of system and the fact that participants had generally not interacted with many other systems like it, it seems likely that while a norm regarding being "nice" or "polite" when interacting with other humans by engaging in non-task talk is highly crystallized, the norm of being "nice" when talking with a non-human other is less crystallized.

The lack of non-task talk among some participants suggests a competing cultural norm that we formulated as:

1. *when interacting with another that is non-human (i.e., the in-car system), if this is done properly, the interaction should be efficient, and one does not need to engage in non-task talk.*

This lack of norm crystallization appears to be present even within the talk of individual participants, such as Participant 16, who uses both "her" and "it" to refer to the system in the statements cited above. He observes, "I don't need her to be my friend," and then later claims, "I just need it to recognize what I tell it to do and do it."

In the case of the second norm, an underlying cultural premise is again involved:

2. *social interaction is based on an agent's (i.e., the system's) status as a human rather than its way of communicating.*

While for some participants the "anthropomorphized" system warrants a certain way of interacting, for others the system is just a tool, regardless of how it speaks, and it does not need to be talked to as if it were a human.

3.6 Implications for speech enabled interface design and suggestions for future research

In this section, we will address the second goal of our analysis—to draw on findings regarding cultural norms and premises formulated above to suggest improvements for future speech enabled technology. First, we propose that designers draw on our findings to make better use of the fact that some users make supportive and evaluative comments. Systems could be designed in the future to monitor for non-task talk and attempt to create a "log of complaints" or "feedback log" that designers, or the system itself, could use to modify the system more largely or tailor the system more specifically to the users' preferences. Designers could thus benefit from the ongoing feedback that some users already provide to the system verbally.

In addition, the prompts and dialog flow of a system could be adapted to respond to a driver's non-task talk by producing non-task utterances—which could create for the driver a sense of having a relationship with his or her system. Developers might even define certain characteristics or personalities for systems, using a variety of voice qualities, and perhaps even give systems personal backgrounds and likes or dislikes, which could further support a driver's treatment of the system as a human interactant. At the same time, developers should also be aware of the competing preference of those drivers who would rather that their system did not respond to or produce non-task utterances; systems should also be developed to accommodate this manner of interacting. The system could monitor user non-task talk and be programmed to produce more non-task talk if the user produced this type of talk, or be programmed not to produce this type of talk if the user did not produce it. The system might also pay attention to those occasions when participants respond (either positively or negatively) to non-task talk that it produces in order to see to what degree and when in the interaction these utterances are appreciated or not.

The competing norms that we have identified here also suggest an area of potential future change, for which designers could prepare. Current drivers, such as our participants, who have not frequently been exposed to this type of technology may be more likely to view "speaking" as indicative of a certain status of agency or personhood. However, in the future, drivers who have more experience with speech enabled systems may tend to view these systems as simply another form of technology, rather than a conversational interactant with whom one might develop a relationship. Developers could work to create flexible systems that might accommodate this type of a change, even for the same user over time. Thus, the system's monitoring of non-task talk would continue over time, and the system might gradually decrease the amount that it produced, if it found that a driver was decreasing the amount of non-task talk that he or she was producing. Alternatively, the system could also gradually produce more if the driver were producing more. This monitoring could also help systems to be more adaptable to different situational or cultural contexts in which different amounts of non-task talk may be deemed appropriate (for example, the situation would likely be different if one were driving his or her spouse in the car versus his or her boss). As mentioned previously, a potential area of future study could focus on how individuals employ interaction with speech enabled technology in order to construct a particular identity for themselves and different types of relationships with other riders in their cars. Findings from this type of research could inform when and to what extent systems are programmed to produce non-task talk.

The ideas we present here may also be relevant for designers working on other forms of technology that make use of a dialogue system, such as smartphones or tablets, and, more frequently, smart homes and smart businesses. Jurafsky and Martin (2009) emphasize the importance of such work on language processing and human-machine interaction when they note:

> The critical connection between language and thought has placed speech and language processing technology at the center of debate over intelligent machines. Furthermore, research on how people interact with complex media indicates that speech and language processing technology will be critical in the development of future technologies.

(p. 15)

As speech enabled technology becomes more prevalent, individuals will increasingly be forced to consider what type of interactional partner a computer represents. Our findings support past research indicating that people will at times respond to a machine that talks as if it were another person, regardless of being aware that the system is a machine. The interactions and perhaps even relationships that those in the future have with their speaking machines may have repercussions in turn for the way in which they understand interactions and relationships with other people, as the quality of being able to speak becomes less exclusively connected with being human and the readiness and willingness of individuals to interact with their computers as people changes.

3.7 Comparison with findings from mainland China

In the data collected in China, we also find instances of non-task talk in the directive sequences. However, only 3 of the 25 participants produced non-task talk when interacting with the system in the data from China, as opposed to 15 participants in the data from the United States. We include here an instance from Participant 23. In this instance, the participant responds to the system's request for him to "please wait 稍等一下。" following a directive to make a phone call by commenting on the system's Chinese language abilities. The participant's utterance in this case was not necessary for the system to complete its task.

Instance 7: Participant 23 - 01:04:40
1 S: Which music would you like to listen to? 您想听什么音乐？
2 P: Make a phone call, to Ms. Zhou 打电话给，给周小姐。
3 S: Please wait 稍等一下。
4 P: Your Chinese is not bad ((laughs)) 你中文不错（笑）

Participant 23 spoke in a low voice, almost a whisper, when stating this utterance. The finding that only 3 Chinese participants produced non-task talk in their interactions with the system would appear to indicate that this type of talk did not serve the same functions as were identified above for the non-task talk produced by participants in the United States and that different cultural norms and premises may have been active in this context. Participants in China who did not produce non-task talk may have viewed the system as a non-human "machine" with whom it was not necessary to be "polite," in a manner similar to those participants in the United States who did not produce non-task talk. Alternatively, participants in China could have viewed the system as "humanish" (like those participants in the United States who produced non-task talk), but not have felt that it was necessary to engage in non-task talk with this type of a "humanish" interactant. Fang and Faure (2011) observe that while Chinese speakers may engage in polite communication with family and friends, with interactants such as "taxi drivers, clerks, shopkeepers, waitresses" "communication is very brief and purely functional" and "there is no greetings and seldom thanks" (p. 328). It seems likely that the system would be treated similarly to a taxi driver or clerk in this situation.

Further analysis would be necessary to determine the norm(s) and premise(s) active in the Chinese context. However, if Chinese participants are indeed interacting with the system as a "humanish" other, who is, however, a "stranger" with whom they do not feel that they need to be "polite," this finding suggests that a further cultural premise is active in the data from the United States. This additional cultural premise would in this case be that not only is the system's way of speaking the basis for how it should be treated, but also, that one should engage in non-task talk with people in this type of a role (i.e. a taxi driver or clerk). Interestingly, Fang and Faure observe that there appears to be a change taking place in Chinese communication in some areas, explaining: "today's market orientation in China has also given rise to the opposite behaviour. For instance, because of increased professionalism, taxi drivers, clerks, shopkeepers, waitresses who traditionally are not polite towards strangers can be very polite today" (p. 328). Thus, the three participants in China who did produce non-task talk may have been influenced by this change— notably, these participants were among the younger participants in the research, being in their early twenties to early thirties. The analysis of the data from China thus highlights a potential additional cultural premise active in the data from the United States.

3.8 Conclusion

In this analysis, we have explored the phenomenon of participants producing non-task talk while engaging in directive sequences. We have proposed two competing cultural norms that we suggest are active in this practice among participants in the

United States. While some participants appear to engage in conduct that could be seen to protect the face of the in-car speech system, others treat the system as a mechanical "tool" with few face concerns, that is merely a device to accomplish a task. We have connected these communicative norms to cultural premises of personhood and differences in how being able to speak is viewed as aligned with an interactant's status. Comparison with findings from mainland China suggest an additional cultural premise active in the data from the United States.

References

Baum, L. F. 1900. The wonderful wizard of oz. George M. Hill Company. United States

Berry, M. 2009. The social and cultural realization of diversity: An interview with Donal Carbaugh. Language and Intercultural Communication, 9:230–241.doi:10.1080/1470847090 3203058

Blum-Kulka, S. 1997. Dinner talk: Cultural patterns of sociability and socialization in family discourse. Lawrence Erlbaum Associates. Mahwah (NJ)

Brown, P. and Levinson, S. 1978. Universals in language usage: Politeness phenomena. In: (E. N. Goody, ed) Questions and politeness: Strategies in social interaction. Cambridge University Press. Cambridge (UK), pp. 56–289.

Carbaugh, D. 1990. Cultural communication in intercultural contact. Lawrence Erlbaum. (NJ)

Carbaugh, D. 2007. Cultural discourse analysis: Communication practices and intercultural encounters. Journal of Intercultural Communication Research 36:167–182. doi:10.1080/ 17475750701737090

Carbaugh, D. 2012. A communication theory of culture. In: (A. Kurylo, ed) Inter/Cultural Communication: Representation and Construction of Culture, Sage. Thousand Oaks, pp. 69-87.

Carbaugh, D., Molina-Markham, E., van Over, B., and U. Winter. 2012. Using communication research for cultural variability in human factor design. In: (N. Stanton, eds) Advances in human aspects of road and rail transportation. CRC Press. Boca Raton, (FL), pp. 176–185.

Carbaugh, D., Winter, U., van Over, B., Molina-Markham, E. and S. Lie. 2013. Cultural analyses of in-car communication. Journal of Applied Communication Research 41(2):195-201.

Dahlbäck, N., Jönsson, A., and Ahrenberg, L. 1993. Wizard of oz studies: Why and how. Proceedings of the 1st international conference on intelligent user interfaces. ACM, pp. 193-200.

Dumas, B., Lalanne, D., and Oviatt, S. 2009. Multimodal interface: A survey of principles, models and frameworks. In: (D. Lalanne and J. Kohlas, eds) Human machine interaction: Lecture notes in computer science. LNCS-Springer-Verlag. Germany, pp. 3-26.

Ervin-Tripp, S. 1976. Is Sybil there? The structure of some American English directives. Language in Society 5(1):25-66.

Ervin-Tripp, S. Guo, J., and Lampert, M. 1990. Politeness and persuasion in children's control acts. Journal of Pragmatics 14:307-331.

Fitch, K. L. 1994. A cross-cultural study of directive sequences and some implications for compliance-gaining research. Communication Monographs 61(3):185-209.

Fraser, B. 1981. On apologizing. In: (F. Coulmas, ed) Conversational Routine: Explorations in Standardized Communication Situations and Prepatterned Speech. pp. 259-272.

Friedman, B. 1997. Human values and the design of computer technology. Cambridge University Press. New York (NY)

Goffman, E. 1959. The presentation of self in everyday life. Doubleday. Garden City (NY)

Goffman, E. 1981. Forms of talk. University of Pennsylvania Press. Philadelphia (PA)

Gordon, C. 2013. Beyond the observer's paradox: The audio-recorder as a resource for the display of identity. Qualitative Research 13(3):299-317.

Grice, H.P. 1975. Logic and Conversation. In: (P. Cole and J.L. Morgan, eds) Syntax and Semantics 3:41-58.

Haddington, P. and Keisanen, T. 2009. Location, mobility and the body as resources in selecting a route. Journal of Pragmatics 41:1938–1961.

Haddington, P. 2010. Turn-taking for turntaking: Mobility, time, and action in the sequential organization of junction negotiations in cars. Research on Language and Social Interaction 43(4):372-400.

Hall, B. 1988/1989. Norms, action, and alignment: A discursive perspective. Research on Language and Social Interaction 22:23-44.

Hall, B. 2005. Among cultures: The challenge of communication. Wadsworth. New York (NY)

Holtzblatt, K. and Beyer, H. R. 2014. Contextual design. In: (M. Soegaard and R. F. Dam, eds) The encyclopedia of human-computer interaction, 2nd ed. The Interaction Design Foundation. Aarhus, Denmark

Hymes, D. 1972. Models for the interaction of language and social life. In: (J. J. Gumperz and D. Hymes, eds) Directions in sociolinguistics: The ethnography of communication. Blackwell. New York, pp.35–71.

Jackson, J. M. 1975. Normative power and conflict potential. Sociological Methods and Research 4: 237-263.

Jokinen, K. and M. McTear. 2009. Spoken dialogue systems. In: (G. Hirst. Morgan and Claypool, eds) Synthesis Lectures on Human Language Technologies, 2(1):1-151.

Jurafsky, D. and J. H. Martin. 2009. Speech and language processing: An introduction to natural language processing, speech recognition, and computational linguistics. 2nd ed. Upper Saddle Prentice-Hall. River (NJ)

Koppel, S., Charlton, J., Kopinathan, C., and Taranto, D. 2011. Are child occupants a significant source of driving distraction? Accident Analysis & Prevention 43(3):1236-1244.

Lakoff, R. 1973. The logic of politeness; or minding your p's and q's. Papers from the Ninth Regional Meeting of the Chicago Linguistic Society. Chicago Linguistic Society, Chicago, pp. 292-305.

Laurier, E. 2005. Searching for a parking space. Intellectica 41-42(2-3):101-115.

Laurier, E. 2011. Driving: Pre-cognition and driving. In: (T. Cresswell and P. Merriman, eds) Geographies of mobilities: Practices, spaces, subjects. Ashgate. Farnham and Burlington, pp. 69-82.

Laurier, E., Brown, B., and Lorimer, H. 2007. Habitable cars: The organisation of collective private transport: Full research report ESRC end of award report, RES-000-23-0758. ESRC. Swindon.

Laurier, E., Brown, B. and Lorimer, H. 2012. What it means to change lanes: Actions, emotions and wayfinding in the family car. Semiotica 191:117-135.

Laurier, E., Lorimer, H., Brown, B., Jones, O., Juhlin, O., Noble, A., Perry, M., Pica, D., Sormani, P., Strebel, I., Swan, L., Taylor A. S., Watts, L., and Weilenmann, A. 2008. Driving and passengering: Notes on the ordinary organisation of car travel. Mobilities 3(1):1-23.

Mäkäräinen, M., Tiitola, J., and Konkka, K. 2001. How cultural needs affect user interface design? In: (M. Reed Little and L. Nigay, eds) Engineering for human-computer interaction Springer. New York, pp. 357-358.

Mao, J., Vredenburg, K., Smith, P., and T. Carey. 2005. The state of user-centered design practice. Communications of the ACM. 48 (3):105-109.

Nass, C. and Brave, S. 2005. Wired for speech. How voice activates and advances the human-computer relationship, MIT Press.

Nass, C. I., and Yen, C. 2010. The man who lied to his laptop: What machines teach us about human relationships. Current. New York (NY)

Oviatt, S. 2003. Advances in robust multimodal interface design. IEEE computer graphics and applications 23(5):62-68.

Passonneau, R.J., Epstein, S.L., Ligorio, T., and J. Gordon. 2011. Embedded wizardry. In: Proceedings of the SIGDIAL 2011 Conference. pp. 248-258.

Pomerantz, A. and Heritage, J. 2013. Preference. In: (J. Sidnell and T. Stivers, eds) Handbook of conversation analysis. Cambridge University Press. Cambridge, pp. 210-228.

Reeves, B. and Nass, C. 1996. The media equation: How people treat computers, television, and new media like real people and places. Cambridge University Press. New York (NY)

Rieser, V., and Lemon, O. 2008. Learning effective multimodal dialogue strategies from wizard-of-oz data: Bootstrapping and evaluation. Proceedings of ACL, pp. 638-646.

Scollo, M. 2011. Cultural approaches to discourse analysis: A theoretical and methodological conversation with special focus on Donal Carbaugh's Cultural Discourse Theory. Journal of Multicultural Discourses, 6: 1–32. doi:10.1080/17447143.2010.536550

Searle, J. 1990. A classification of illocutionary acts. In: (D. Carbaugh, ed) Cultural communication and intercultural contact. Lawrence Erlbaum. Hillsdale, pp. 349-372.

Shahrokhi, M., and Bidabadi, F. S. 2013. An overview of politeness theories: Current status, future orientations. American Journal of Linguistics 2(2):17-27.

Tannen, D. 2004. Talking the dog: Framing pets as interactional resources in family discourse. Research on Language and Social Interaction 37(4):399-420.

Tsimhoni, O., Winter, U., and Grost, T. 2009. Cultural considerations for the design of automotive speech applications. In: Proceedings of the 17th World Congress on Ergonomics IEA 2009, Beijing, China.

Turkle, S. 2011. Alone together: Why we expect more from technology and less from each other. Basic Books. New York (NY)

Winschiers, H., & Fendler, J. 2007. Assumptions considered harmful: The need to redefineusability. In: (N. Aykin, ed) Usability and internationalization, Part I. Lawrence Erlbaum Associates, Mahwah, pp. 452-461.

Winter, U., Shmueli, Y., and T. Grost. 2013. Interaction styles in use of automotive interfaces. In: Proceedings of the Afeka AVIOS 2013 Speech Processing Conference, Tel Aviv, Israel.

Winter, U., Tsimhoni, O., and T. Grost.2011. Identifying cultural aspects in use of in-vehicle speech applications. Paper presented at the Afeka AVIOS Speech Processing conference, Tel Aviv, Israel.

4 User Interaction Styles

When we decide to interact verbally with a machine a variety of questions are immediately raised about how this should be done. One central aspect of this decision is informed, in part, by the relational implications, or lack thereof, presumed by engaging in that kind of interaction. Is this thing I am talking to something like a friend, entertainer, secretary, or, simply a machine? One could imagine that depending on the answer to that question a differing set of interactional styles may be employed that both reflect and enact this orientation. The use of these styles should not be understood as invariably attached to particular persons or person-types, but rather, are employed in ways that reflect participants' emerging and ongoing goals for the interaction and the social meanings that are expressed through the use of those styles during particular events and situations.

In what follows we identify two primary interaction styles enacted by users with an in-car communication system. The particular features of each style in-use are discussed, as well as the meta-discourse that users produce when speaking about their use of that style and their more general relational orientation to the system. We then compare the use of these styles cross-culturally drawing from data gathered in Beijing and Shanghai. We conclude with a discussion of implications for system design.

4.1 Dimensions of Stylistic Variation

We began the process of identifying stylistic variation through the analysis of 596 instances of task events across 22 participants. Task events in the corpus were bound by participant's initiating an interaction with the system in order to ultimately accomplish a particular task, like tuning the radio, making a phone call, playing their personal music catalogue, or navigating to a destination. Our analysis revealed a set of dimensions along which participant's engagement with the system varied. These include variation in a recurring preference for:

1. Greater or fewer number of conversational turn exchanges relative to the minimum number of requisite turns for task accomplishment.
2. Elaborateness of utterance construction relative to the shortest utterance possible for task accomplishment
3. Use of politeness indicated by presence of "please" "thank you" "can you" etc.
4. Speech act formulation including a relative preference for phatic expressives, directives, or inquisitive formulations
5. Performance additions including various discourse markers and exclamation tokens like "huh" "uh" "um"

https://doi.org/10.1515/9783110519006-004

These dimensions of stylistic variation are apparent in the instances below and exemplify how these preferences may cluster into styles, though they need not necessarily do so.

4.1.1 Dimensions of an Efficient Task Accomplishment Style

Instance 1
Participant 6 - 01:29:01.60
P = Participant, S = System
1 P: ((touches microphone button and system dings)) Play artist
2   Creedence Clearwater.
3 S: Playing Chronicle volume 1, select a song if you want.

In this instance the participant accomplishes a desired task, playing music from a particular artist, in a highly efficient and minimally interactive form. The task is accomplished as a "one shot utterance" where the participant provides enough information for the system to execute the command, with nothing extraneous to the goal of task accomplishment included. Note, for example, that the utterance begins with an action verb "play," rather than a modal verb "can," as in "can you play the artist Creedence Clearwater," which might indicate a preference for shaping directives as inquisitives. Also absent are potential politeness markers, as in, "can you please play Creedence Clearwater?" Such a phrasing avoids the use of short explicit directives, opting instead for shaping the directive as an inquisitive, which coupled with the use of the politeness token "please" can function to yield some humanistic agency to the system and legitimize its treatment as an interactional partner that merits a respectful interactional orientation. The formulation also includes no markers of hesitation, exclamation, hedges or softeners as might be seen in the case of "uhh, can you please play Creedence Clearwater?" The following instances further demonstrate the potential for system interactions to be approached by users in highly efficient ways.

Instance 2
Participant 20 - 21:34.74
1 P: ((participant touches microphone button and system dings))
2 S: What sports channel do you want to hear?
3 P: Go to NPR.
4 S: Hold on. ((pause)) Tuning radio to 88.5 FM WFCR.
5   ((radio station changes))

Instance 3

Participant 12 - 01:11:45.40

1 P: ((participant touches microphone button and system dings))
2 S: What XM channel?
3 P: Call H-.
4 S: Got it. ((radio stops playing and phone rings)) calling H- at home

This preference among some users, at some times, for accomplishing tasks in the most minimal number of turns necessary, with utterances formulated as shortest possible directives, and absent of any indicators of politeness, human social concerns, or speech perturbations, is here conceptualized as a style preference of "*maximal efficiency*."

The highly efficient approach to task accomplishment apparent in the above instances are notable in their absence of other conversational features that might be present, but here are not, as in the hypothetical reformulation of "uhh, can you please play Creedence Clearwater?" However, not all users make use of this style of *maximal efficiency*, instead opting to accomplish tasks in a more *interactive* fashion, as seen in the following instance.

4.1.2 Dimensions of an Interactive Task Accomplishment Style

Instance 4

Participant 13 - 01:18.85

1 P: ((touches microphone button and system dings)) Play XFM.
2   ((pause))
3 P: No you don't like that one?
4 S: Please wait.
5 P: Ah
6   ((pause))
7 S: Please review your station it may not be valid.
8 P: ((laughs)) Ok mm. ((pause)) Play my music.
9 S: Please let me know what you want.
10 P: Ok. You are a goofball. I'm going to call you-
11 S: Wait a moment. What kind of music would you like to hear?
12 P: My music. You can do it. Come on.
13 S: What song do you want to hear?
14 P: Hmm. Play Creep.

15 S: Hold on.
16 P: You can do it baby.
17 S: Playing Creep. ((music plays))
18 P: There you go.

In this instance, while the participant's initial directive to the system is formulated in a short, efficient declarative utterance, "play XFM" on line 1, the participant elects not to simply wait for the system to respond, but rather engages in a kind of playful conversational banter saying "no, you don't like that one? When the system confirms it cannot understand or find the station that has been requested it asks the participant to reissue or reformulate the directive so that the system may attempt to comply again. This prompts a laugh from the participant on line 8, followed by a brief confirmation that the user understands and will play along, but rather than silently formulating a new directive the participant deploys the discourse marker "mm" as a verbal performance of thinking that the system cannot understand or act upon. The participant proceeds with a pause that prompts the system to follow up with the participant saying, "please let me know what you want," on line 9.

The participant appears to orient to this follow up request by the system as a mishearing of the directive to "Play my music" and offers an assessment of the system as a "goofball" and begins to issue a personal name for the system on line 10 saying "I'm going to call you-" before the system cuts in.

While the participant begins this task request with a short "one shot" utterance that includes only information necessary to task accomplishment, they quickly move to an interactive style marked by verbal embellishment, elaborate turn construction, and interpersonal social features. Ironically, that very move to an elaborate interactional style introduces a host of supplemental talk into the interaction that the system is not equipped to monitor for, understand, or act on, and as a result produces a series of misalignments between the participant and system, beginning with the introduction of "no, you don't like that one" on line 3, that is ultimately resolved after a 16 turn exchange.

In this instance, the participant's talk is marked by non-task related talk, some of which implies an interpersonal relational orientation to the machine, as in the participant's repeated statements of encouragement and support "you can do it baby," and the ultimate congratulatory affirmation "there you go" on line 18. While this style may not maximize or emphasize the possible efficiency of human-machine interaction, it does serve to highlight the potential desire for achieving humanistic interpersonal relationships with machines as an end unto itself, accomplished in this instance through attempts at generally friendly conversational dialog at the expense of efficient task completion.

This stylistic orientation to the machine as a potential humanistic interactional partner, characterized by playful and non-task oriented talk, where the accomplishment of the task unfolds over a series of turn-exchanges and featuring various discourse markers like "ah" (line 5) "mm" (line 8) and "hmm" (line 14) is here conceptualized as a style of "*maximal interactivity.*" Unlike the maximally efficient style evident in the first 3 instances reviewed, which orients to the system as a simple machine, this style orients to the system as a humanistic-agent, and invites the system to engage in humanistic practices of being fun, playful, and entertaining. In the instances below the features of this style are further demonstrated.

Instance 5

Participant 9 - 53:32.70

1 P: ((touches microphone button and system dings)) Give me WFM-
2   uhhhh (.) give me W:: what's it called FCR
3 S: Pardon?
4 P: FCR
5 S: Please confirm 88.5 FM WFCR
6 P: Good. ((pause)) Good job ((shows thumb up))
7 S: Tuning radio to 88.5 FM WFCR ((radio station plays))

Instance 6

Participant 24 - 42:17.40

1 P: ((touches microphone button and system dings))
2   (4.0)
3 S: Pardon?
4 P: Oh, so you're waiting for me to talk? ((Laughs)) OK. How about can
5   I have My Music please?
6   (5.0)
7 S: Could you repeat that please?
8 P: My Music please?
9 S: What kind of music would you like to hear?
10 P: Albums
11 S: What album would you like?
12 P: umm. 21?
13   (5.0)
14 S: Playing 21 or select a song.

In these instances, we note the persistent presence of talk that is not required in order to for the system to accomplish the task requested by the user, and which, at times, may even interfere with or complicate that goal. Here, we conceptualize this as "non-task talk," which is reviewed in detail in chapter 3 and elsewhere (Molina-Markham, et. al. 2016). The inclusion of non-task talk is evident in each participant's turn exchanges, as in participant 13 on line 3, ("No you don't like that one?"), participant 9 line 6 ("Good. ((pause)) Good job ((shows thumb up))"), and participant 24 line 4 ("Oh, so you're waiting for me to talk? ((Laughs))"). In chapter 3, we suggest the inclusion of non-task talk may have implications for relational development with the system and often reflects an interactional norm premised on an expressed orientation to the system's limited rights to a kind of personhood, as well as accompanying premises about the appropriate relational alignment between human and machine interactants as relatively human-like.

In addition to non-task talk however, the instances above also exhibit the unfolding and incremental quality of task events making use of an interactive style. While participants who employ an efficient style will often generate a directive in "one shot" where all requisite information is condensed into a single utterance "play Creedence Clearwater," participants making use of an interactive style may incrementally release this information through a series of turn exchanges as in participant 24 who first directs the system to play "My Music" then to play an unspecified "Album" and then finally selects "21." This might have been accomplished through the directive "Play Album 21," but in this case is done through the interactive quality of question-answer sequences that prolong the interaction in a style that more closely mimics the given and take turn-exchange of interpersonal relations, rather than the socially distant and command-oriented style of machine interaction.

Like participant 13 above, these participants also make use of politeness markers, such as "please" (instance 6 line 5 and 7) or compose their utterance as an inquisitive rather than a directive, using a wider range of speech acts (instance 6 line 4 "How about can I have My Music please?" or on line 7 saying "could you." Participants 9 and 24, like 13, also exhibit speech that has not been composed to eliminate various discourse markers like "umm," "uhhh" or exclamatory tokens like "Oh." This suggests an orientation on the part of these users to the production of a directive as an evolving and exploratory interactive process to be negotiated and accomplished with the machine as a kind of partner, where being surprised "Oh" or unprepared "uhhh" is all part of the process of getting to completion. Efficient users however betray an alternate orientation to interaction with the system that privileges product rather than process, where knowing in advance what you are going to say, and having formulated the directive in a maximally efficient way, assures the shortest path to task completion.

4.1.3 Distribution of Styles

While we have presented 3 instances of each style here, use of efficient and interactive styles were not evenly distributed across participants. Among the 22 participants analyzed here, 3 exhibited a strong preference for a more elaborate interactional style, and 3 with moderate use of an interactional style, who together produced 78 instances of interactively styled task events. The remaining 16 participants exhibited a preference for a more efficient style, with some occasionally using a more interactive style. These 16 participants produced 518 such efficiently styled task events.

All task events wherein a style of maximal efficiency was employed also tend toward shorter interaction sequences and minimized number of turns. In almost 55% of these events (281 out of 518) the user provides all necessary information for task completion in the first utterance, a "one-shot," as can be seen in the examples from participants 6, 12, and 20 above. Users who took two or more turns to provide all necessary information to the system mostly used an incremental strategy of moving through the interactive question and answer sequence, each time providing minimal new additional information until the system had enough information to accomplish the task. One might imagine that a user would employ such a process because he was less familiar with the possibilities or capabilities of the system, potentially unaware that all necessary information could be provided to the system in "one-shot."

However, in task events where the participant distributed the information over three or more turns, 53% (41 out of 78) of the time this occurred even after becoming familiar with the system and its functionality. This suggests that participants who employ a more interactive style do so not because of a functional incompetence but because they prefer to. In order to better understand how participants who employ a more efficient or interactive task accomplishment style understand and account for their own behavior we now turn to an analysis of participants' own talk about their stylistic preferences for interacting with the system. We further analyze this talk for a set of cultural premises about personhood, social relations, emotion, and communication itself that inform the production and interpretation of this way of speaking to the system.

4.2 Stylistic Preferences in Verbal Reports

Over the course of the study interviews were conducted with each participant, during which participants expressed their verbal interpretations of the behaviour of the system, as well as their future desires for the interactional quality of the system. Often these expressed desires clustered among participants along their usage patterns with those employing a style of maximal efficiency expressing a set of desires for the system's ideal behaviour, and with those preferring a style of maximal interactivity expressing a different set.

Participants who often interacted with the system employing a style of maximal efficiency reported that it was most important to them that the system provide an effortless, easy and seamless interaction that was consistent, short and efficient. Participants further expressed an understanding that this preference for smooth efficient interaction may require routinized interactional patterns that require them to adapt to the system's verbal command structure. This can be found in the report of participant 14 who says "I don't mind knowing specific commands, you need specific commands for everything that you do, whether they're verbal or anything else. That's not really something that bothers me. But just being... yeah, just being able to tell it what you want and not have to like poke something or do something. I think that's the most useful for me that it would be".

Other participants who employed a style of maximal efficiency focused on their desire for consistency, predictability, and speed, as in "Some <communication events> are very smooth, and you can just talk right away. Other ones... the lag between when it will prompt you for what it wants varies, it seems considerably like a second or two at least. The spelling up the radio station designation every time it switches the channel can get old." This expressed desire for efficient communication can also be found in the report of participant 4 who positively evaluates it's "efficient" performance, confirming an implicit premise that the efficiency of the system can and ought be a major evaluative criteria saying, "I think it's pretty efficient overall, with the questions, you know and confirming that is fine". Participant 20 negatively evaluates the system's efficiency for its perceived lack of consistency and excessive information offering, which they hold responsible for overly elongating turn exchanges unnecessarily -- "uhhmm, a little bit inconsistent. I feel like... uuhhmm... you know, it's definitely.... I think that... in the best interest of the company trying to do this, the faster that it responds to you and processes what you're trying to get it to do, the better it's gonna be. 'cause, I mean, people who have a short attention span, they don't want to wait for things." Participant 17 expresses a similar concern about whether the commands could be shortened, "I would also think that some of those commands could be quite a bit shorter ( ) 'which XM station or channel do you want to hear?', that's almost TMI".

Participants' sense that the system could be further stripped down only to its most basic and necessary interactional functionality, even at the expense of more closely mimicking human interaction is captured in participant 16's report saying "yeah... I don't need this ((points to the screen)) to be human, I just need it to recognize what I tell it to do and do it. I don't need any of the 'ok, see you later', you know, I don't take any comfort from it having a couple of pre-programmed personable phrases". For these participants, their expressed desire for the system appears entangled in their premises about what a system like this can and should be. A system that exists solely to accomplish task directions through a verbal channel so as to free the driver to focus on the road may not "take any comfort" from its attempts at human approximation. However, other participants did express an imagined role for the system as more than a task completion engine, orienting more to the system as a potential interactional partner.

Participants who predominantly employed a style of maximal interactivity often expressed a preference for being able to speak to the system in conversational ways, unrestricted by the confines of a rigid verbal command structure. Participant 13 captured this desire saying "it would be nicer if it kind of understood my conversational style well enough" and "I mean I would treat it like a person. I'm generally nice and polite but then if it was being smart-ass I might yell back at it, but I wouldn't, you know. I wouldn't get mad, not really." Participant 24 also expressed a preference for the system to employ a more humanistic conversational style saying "you feel like you're interacting with somebody and having a conversation", and "that's something that I just feel better doing that, I have no idea why. It's very strange. When I think about it, it just feels more natural."

Participants also expressed their sense that part of a more humanistic conversational exchange may include the use politeness markers, as in participant 13 who said "Well it's technology. That doesn't mean you don't have to be polite." Participants also sometimes evaluated the system negatively for being impolite or engaging in behaviour that could be interpreted as rude or inconsiderate of their interactional partner as in the case of participant 9 who criticizes the system for its impolite turn-exchanges -- "It's not listening carefully. It needs to pause and wait for his turn to speak. Even if it's just saying 'I didn't understand you'." Sometimes the inclusion of a politeness token like "please" was cause enough for a positive evaluation, as in participant 7 who likes the system "'cause SHE said please to ME, which I thought that was very nice."

These same participants further express an aversion to precisely the kind of interactional style desired by users who employ a style of maximal efficiency. When asked how it feels to have to give the system orders, she said "Yeah, I don't really like that." Participant 24 echoed this preference for polite natural language over bold directives, "... like you would be speaking to someone, it just feels more natural for me to have etiquette and be polite."

These reports implicate an orientation to the system as an interactional partner with whom a humanistic relationship is generated through certain kinds of conversational engagement, and further that this type of relationship is preferred and more "natural". This orientation to the system as a kind of pseudo-person is expressed by participant 19 who explains "it's talking to you, so I think ( ) to treat it more like a person." Participant 24 elaborates the potential relational benefits of this orientation saying, "It's kind of fun to have someone to chat with, to see what it would say, to test it out. Or to remind me. I like that personal feature ...you feel like you're interacting with somebody and having a conversation." In this formulation the system acts as a kind of buddy to "chat" with, and as participant 9 suggests, might even extend to political debate, saying "Well if I say 'I wanna hear some wacko political rhetoric' and you don't know what that means... and of course as soon as I hear that stuff I wanna talk about it with somebody so... it's tempting not to wanna talk back to it".

Overall, the interview data suggest that multiple norms of interaction are at work in expressed preferences for the progression of the act sequence, the interactional/relational goals that participants believe can and should be accomplished through these sequences, and the relationships between users and a machine that participants believe should or should not exist in this context. In the following section we move to draw out more explicitly the features of each style-in-use as well as participants' interpretations of how system behaviour can better meet the range of stylistic preferences.

4.3 A Style of Maximal Efficiency

As discussed above, certain discursive features emerge in interaction between participants and the system. When clustered together, they constitute a style-in-use of Maximal Efficiency. In their most extreme formulation these include:
-   Preference for task initiation performed in "one-shot" (e.g., "play radio 88.5"),
-   Preference for task switching performed in "one-shot" (e.g., "now call John Smith at home"),
    o   User comprehension of the system's ability to process requests formulated in this way may be a limiting factor here.
-   Preference for whatever method the user perceives and reports as most efficient, which may sometimes include an incremental approach to task completion, where the task is distributed across a number of turns, *if* doing so is believed to be more efficient. (This may occur when the user is uncertain about the speech systems capabilities or is exploring for the purpose of content discovery, as in browsing available music).
-   Preference for directives formulated as commands, with repeatable structure.
-   Preference for most minimal elaboration of directive formulations.
-   Preference for task endings through one-shot voice commands or touch of the "end" button.
-   Preference for minimal number of turn exchanges and overall interaction time.

Users who employ a style of Maximal Efficiency expect the system to match their style and assess the system's performance on criteria relevant to their preference for an efficient interaction.
-   Prefers system to minimize number of turns and overall interaction time.
-   Prefers system to play short prompts, with no relational features.
-   Prefers system to provide feedback only when information speeds current or future task completion.
-   Prefers system to be designed toward high task completion rate, and is indifferent to, or negatively assesses, system attempts at politeness, humanistic voice, humor, or other features that would imply or seek to establish human interpersonal relations.

4.4 A Style of Maximal Interactivity

While the above configuration of preferences as both observed in, and reported about, user interactions with the system is a prevalent orientation to interacting with and assessing the system, an alternative orientation that seeks to maximize the humanistic interactivity of the system can also be identified.

Participants who employ this style report enjoying playing with it, calling it a name, inviting and desiring a response. While the primary goal for these users is task completion, as for predominantly efficient users, the interactional path there embraces the journey, as well as reaching the destination, as part of the satisfaction of using the system. The features of this style configure, in an extreme form, in this way:

- Preference for task initiation including non-task related talk (e.g., "well I wonder what I am in the mood for right now..."),
- Preference for including non-task talk in task switching events (e.g., "well, should I call Bill now or Janice...").
- Preference for addressing the system with an ascribed name (e.g., "OK Eliott, what's it gonna be now...") or with an endearment term (e.g., "nice job Darlin'").
- Preference for non-repetitive, distinctive and varied directive formulations
- Preference for task completion accomplished in an incremental form, with requisite task completion information potentially spread across multiple turns. (e.g., "how about some country"; "do we have any Garth Brooks available.."; "is his latest CD in there..."; "what about the duet with Patty Loveless?" and so on)
- Preference for politeness in directive formulation, including softeners and other face-saving devices. This preference extends to task completion as well including terminal "thanks" and "see you again."

Users who employ a style of Maximal Interactivity expect the system to match their style and assess the system's performance on criteria relevant to their preference for an interactive dialogue sequence.

- Prefers system to express some "personality." The features of this personality are ideally configurable by user.
- Prefers system to provide more elaborate feedback not only for the goal of efficient task completion, but for pleasant interaction, such as backchannels for listing and understanding, apologies, and politeness phrases, address terms, and even some non-task talk which mirrors the type of relationship that the user wants to establish.
- Prefers the system to be informal and polite.
- Prefers the system to be able to react to utterances other than related to the predefined number of possible tasks.

– Prefers the system to prioritize being "nice" and "helpful" over reduced number of dialogue turns/overall interaction time.

4.5 Contextual Dependence

The data as well as the interviews indicate that these interaction styles are not directly related to types of people, but to different ways of using the system at different times. Among the 22 users analysed, the 3 users with strong preference for interactive style occasionally became efficient in their utterances, the 3 users with moderate use of interactive style varied from efficiency to interactivity when performing their tasks. Out of the remaining 16 predominantly efficient users, 7 users had communication events with more elaborate utterances using some characteristics of the interactive style. In this way no user commits in advance for all time to the use of a particular interactive style. Interaction styles are more accurately conceived as a continuous axis from maximal efficiency to maximal interactivity with each user predominantly situated in one area on this axis and with occasional claims to other areas. Users may employ varying styles from moment to moment depending on the driving context, situation, time, mood, etc. Users may vary from their preferred style in a certain context, such as a daily commute versus a weekend trip with the family. When thinking about the system's ability to adapt its style to varying user desires participant 9 mentioned that he would enjoy having both styles as available modes to choose what suits his mood, the situation, task, etc.

While the data clearly display a tendency for contextual dependency in the observations as well as in some interviews, the field study did not provide sufficient drive time, nor context variation to determine with confidence under what set of conditions a user may adopt a particular style over another.

4.6 Cultural Interaction Styles in China

Similar to US participants, Chinese participants in the study predominantly used the efficient style with 17 out of 25 participants preferring an efficient style. However, while no significant difference was found in style selection on the basis of gender in the US, the Chinese study found that female participants were more likely to use the interactive style than male participants. Even after being prompted by the researcher to try a more interactive style more often, only 6.6% of male users utilized it compared to 22.1% of female users. This finding indicates the possible existence of differences in Chinese norms for gendered communication, and that these norms may also govern interaction with non-human systems. More research needs to be conducted with regards to this phenomenon.

Chinese users who employ an efficient style shaped directives in a grammatically and linguistically minimal way, with a preference for concise commands, for example saying "停" for "stop" rather than "结束音乐" for "end music":

Instance 7
Beijing Participant 6
1 P: (Touch microphone button)
2 P: <u>Stop</u>   停
3 S: Which music station would You (polite form) like to listen to? 您想
4   听什么音乐类电台?
5 P: <u>Stop</u>   停

While Chapter 6 details the pattern more thoroughly, we note here that user-initiated speech overlap is noticeably more prominent among Chinese participants than US participants, with Chinese participants being more willing to initiate a new command, or repeat what may have been understood as an unheard command, while the system was still speaking. At times this may have been done in order to terminate an ongoing or preliminary course of action on the part of the system, as a user clarifies or alters their desire, as in the following:

Instance 8
1 P: (Touch microphone button)
2 S: You would like to listen to what music? 您想听什么音乐？
3 P: Liu Ruoying's (an artist name) 刘若英的
4 S: Play Liu Ruoying [or select one album. 播放刘若英[
5   或选择一张专辑。
6 P: [Don't want it. I would like to listen to another one. [不要
       了，不要了，我要听

Some Chinese participants who employed a more incremental approach to task completion sometimes reported doing so because they were uncertain whether the system understood their commands, and accounted for variation in their approach to task completion on the basis of their perception of the system's capacity for language processing. Shanghai participant 11, for example, used a step-by-step approach and later switched exclusively to a one-shot approach once he concluded the system could understand his commands:

> If the speech recognition is accurate enough, which means it could recognize fuzzy, unclear commands, I prefer the one-shot approach. It's faster. If it could not understand me well, the step-by-step approach is better. Again, it's about efficiency. If the system is smarter, (and) could recognize unclear commands, then (I) use the one-shot approach. If not, then (I) use step-by-step.

这个需要看的，如果我说的话它识别率非常高的话，就是说它应该有一种模糊的识别能力，如果能这样的话，我倾向于直接的命令，比较快一点。但是如果它不能理解我的话，还是菜单式进入好一点。

However, some Chinese participants preferred an incremental approach to task completion even after discovering the capabilities of the system. These users also preferred indirect requests instead of commands, and sometimes used humour when interacting with the system -- some key dimensions of the interactive style as identified in the US context above. Users who employed an interactive style in the Chinese study also used the polite term of address 您 (nin), when addressing the system, incorporating some features of talk that carry primarily relational meanings.

While a consistent feature of the interactive style in US contexts was the use of non-task related talk during the issuing of directives, among other times, Chinese participants whose talk features other markers of interactive style, engaged in far less non-task related talk, as reported by Chinese members of our research team. However, it is possible that this analysis is itself subject to variation around what constitutes talk that is considered "task related." As a result, we believe this finding requires further investigation and sensitivity to cultural variation surrounding theoretical constructs such as these, not only among those who use these systems, but those who analyse them as well.

In interviews, Chinese participants sometimes evoked social identity categories in their conceptualization of the system-as-interactant. These included repeated references to the system as a kind of assistant or service worker.

Beijing participant 1

> It is like an *assistant (emphasis added)*, from whom I could get a lot of information and could help me do many things. I want it to be my *staff*.
> 像助手，助理一样，你能从它那儿获得很多信息，它能帮你做很多事情。我希望它是我的下属。

Beijing participant #11

> It is like my *secretary*, helping me do many things, simplifying the tasks, and also keep me safe.
> 它就相当于一个秘书一样，帮助我做很多事情，简化很复杂做的事情，保证我的行车安全。

As illustrated in these examples, key terms such as "assistant", "staff," and "secretary" indicate a perceived hierarchy between participants and the system invoking a service orientation to the relationship. Messages that shaped the system in this way were not as present in U.S. American data, where the system was mostly shaped in discourse as a partner. As was noted in Chapter 5, small talk with service workers is not a common practice in daily Chinese social interaction, and as a result may help account for the diminished role of non-task talk in the Chinese data.

The orientation to the system as a worker in service to the user may also be iden-tifiable in the following participant's lament that the system failed to check in be-fore ending the system on whether the user had any further needs.

Beijing participant 3

> After playing the song, there was no other music and the system did not ask me *"anything else"* before closing the application. This needs to be improved.
> 歌之王放完就完了，没有后续歌曲，也没有问"还有其他指令吗"等。放完即冷场。

The participant's shaping the system's programming to end the task once it is com-pleted without a follow up offer for further assistance as a design flaw may be con-nected to cultural premises about the appropriate relationship between a competent service worker or assistant and their "boss," where cultural norms prefer that the employee should ensure a person's need is fully met prior to him/her closing the interaction.

A competent service worker or assistant would also be able to put the customer or boss at ease by displaying a friendly yet polite demeanour. Although it is evident that Chinese participants shape their relationship to the system as hierarchical, participants shared how they prefer the system to be "friendly yet respectful," to do away with formality while never neglecting their vertical relationship. As Beijing participant #7 remarked:

> (The system) does whatever I want, either shuffling or playing a specific song. It chats and talks with me especially during a long drive. So the system makes me feel well, not alone. *Friendly, but respectful (emphasis added).*
> 你要说播放，就是随机播放，要是挑，它就播你挑的。能聊天，能跟它说话，尤其长途驾驶的时候，语音系统比什么都舒服。就是不孤独了。

This expressed desire that the system should be both friendly but also respectful in a way that demonstrates appropriate deference to the hierarchically superior rela-tional position of the user is distinctive in the Chinese corpus, as the use of US American styles with the system often positions it either as a relationally distant "machine" where mutual respect is not presumed important or as a kind of friendly partner to chat with where the expression of deference or invocation of hierarchy would only serve to highlight the non-egalitarian nature of the system, thereby de-stroying the carefully crafted space of suspended disbelief that allows users to play at humanistic interaction with machines.

As a result the use of a predominantly efficient style in Chinese interactions should not be assumed to be indicative of an orientation to the relationship with the system as a non-human other, as is often the case in US uses of efficient style, as Chinese cultural preferences for this style may be connected to their appropriate use with human service workers.

4.7 Considerations for In-car HMI Design Recommendations

As people use their in-car communication system, at times they prefer a style of maximal efficiency. When they do, they want to minimize the time it takes for dialogue and the number of turns it takes to complete a task. The implications for interaction design in support of this style are as follows: create a possibility for the interaction flow to be as concise and efficient as possible, subordinate prompt design to this optimization principle, remove unnecessary prompts or other elements in the dialogue process, and in turn taking. When the maximal efficiency protocol is active, users find the absence of interactive elements to be acceptable and even pleasant. And in turn, when the optimal efficiency style is preferred, the interactive elements in the prompts are treated at best indifferently or at times as irritating.

Users at other times prefer a style of maximal interactivity. When this is the case, users show a preference for variation, relational talk, naturalness, and being entertained. At these times, users appear to want to develop a humanistic relationship with the system. The implications for interaction design in this case are that prompts should be designed to approximate a human-human interaction and include regular features of friendly social interaction including politeness. Thus, prompts for this situation should be designed to progress toward task completion, and provide opportunities to respond to user non-task related talk.

Because users may not know if advance the kind of style they want to employ, how they would like the system to respond, or under what conditions they may want to employ a different style, we conclude that a system that can determine user interaction style and adapt on an ongoing basis would be ideal. A set of system prompts could be designed that accommodate each of the two styles that we have identified here. With this type of a system, users could select which style they prefer and could also potentially switch between styles. One way of accommodating this switching would be to first design prompts emphasizing the efficiency principle and then develop these prompts by adding interactive elements, so that the basic prompt is the same for both sets of prompts.

Implications for system design in the Chinese data are informed by the hierarchical manner in which Chinese participants perceive themselves to be related to the system. Following this superior-subordinate relational pattern, we recommend that the opening speech prompt be delayed to allow for the user to take control of turn-taking and topical choices in his/her interaction with the system. In other words, the system should allow the user to decide when to talk, the topic of the dialog, the dialog style, and so on. The system should listen to him/her and respond to selected topics only afterwards.

One way to solve this particular issue is to provide a quick tutorial for the user to accelerate his/her fluency in communicating with the system. The system could give the following prompt: "please talk after the audio ding." This prompt could then be removed after the learning period is over. Knowing that the audio ding signifies turn taking, the user could exercise his/her discursive power in determining the conversational flow, including turn taking sequences and topics of interest, in his/her interaction with the system.

4.8 Summary

One finding from our field studies includes these two different styles--one with a focus on efficiency in task completion and a second that emphasizes elaborate interactivity and a personal relationship with the system--that are preferred by participants to varying degrees in a northeast region of the United States. We add at this point that our field data collection in China, and its comparative analyses, revealed the same two general interaction styles, with differences between the two mainly being in the language itself including: 1) the dialogue flow relative to its sequential structuring, and 2) the introduction of features which show respect from the system-as-a-subordinate to its user. This reveals differences in directive formulation, error corrections, and event structure generally. Using these styles as guides, among other culturally distinct features (see this research teams' other published field studies), researchers can develop speech enabled systems that more closely align with user expectations and preferences, here demonstrated briefly to illustrate interaction flows and prompt design. In terms of future work, the field data we have collected is invaluable, deeply rich and detailed enough to facilitate the development of the exact phrasing of prompts, the detailed specification of the voice persona for both interaction styles, the flow and rhythm of the system, as well as development of the other design dimensions discussed above.

References

Molina-Markham, E., van Over, B., Lie, S. and D. Carbaugh. 2016. "You can do it baby": Cultural norms of directive sequences with an in-car speech system. Communication Quarterly 64(3):324-347.

5 Taking Turns with a Machine: The cultural maintenance of interaction organization

5.1 Introduction

Imagine a conversation wherein you ask a friend if they would like to see a movie. The friend replies "what concert did you want to see?" Your first thought might be, "What? How is that relevant to the question I just posed? Why might they have said that? Did they mishear me? Perhaps they were implying they would rather go to a concert than a movie? Humans routinely use this kind of answer, one that flouts Grice's (1975) maxim of relevance to accomplish communicative goals like conversational implicature. However, what are we to make of the following interaction between a human user and an in-car multimodal infotainment system?

P = Participant, S = System
1 P: (Participant touches microphone button)
2 S: (audible ding)
3   (0.6)
4 P: phone ca[ll
5 S: [which station or channel do you want to hear?

Here we see much the same oddity as in the hypothetical conversation between friends; the user asks to make a phone call and the system replies with a question about what radio station they would like to hear. Except unlike our conversation between friends, here, the user is not likely to wonder if the system would prefer to listen to a radio station or might be attempting a conversational implicature. So how, as a user of a system like this, do we make sense of this interaction when the framework we use to interpret similar human speech no longer works? How did we get into this situation to begin with, and what do we do about it now that we are here? Finally, what role might variation in users' cultural premises play in informing these interpretations?

Users of multimodal systems like this are often faced with these kinds of communicative challenges because machines are imperfect interactants and frequently fail to follow basic rules and principles for the governance of communication that human interactants generally follow with each other. That some of these rules vary in their application and meaning cross-culturally only further complicates matters. This poses a problem for studying these human-machine interactions through the application of many existing theories of social interaction because of their reliance on the assumption of a "model interactant" who is competent in the culturally distinctive ways humans have developed for communicating with one-another. This model interactant follows particular rules for the organization and ongoing management of

https://doi.org/10.1515/9783110519006-005

conversation that are based on the assumption that humans interact from a set of what some have suggested are universal principles (Sacks, 1974; Sidnell, 2001; Stivers et al., 2009). However, machines in interaction with humans routinely violate behaviors expected of this "model interactant." This is because machines do not currently exhibit the kind of rationality presumed by some theories of communication (Brown & Levinson, 1987; Grice, 1975) nor share the "face" (Goffman, 1959) concerns about the impact of their communicative behavior on their social identity, which informs much of human interaction. They may seek to operate cooperatively (Grice, 1975), if programmed to attempt to do so, but may also violate common interactional practices in obvious ways, which when done by human interactants with each other are often taken to be meaningful violations in and of themselves, since a rational cooperative human interactant would likely only do such a thing intentionally. In the case of human-machine interactions, however, the machine has no intention of implying some meaning through its violation of interactional norms, and users likely know that.

This means that some of our fundamental assumptions about social interaction become unreliable in human-machine interaction. And by extension, machines themselves may be found to be unreliable interactants for these very reasons. Two questions then become essential for us to pursue in order to better understand the dynamics at work in human-machine multimodal interactions: 1) how do humans interact with machines that are not assumed to be, nor able to operate as, fully culturally competent interlocutors? and 2) how do we manage moments when things inevitably go wrong in these interactions?

One area designers of such systems have often overlooked in their attempts to create ever more human-like interactants is the structure of turn exchange in conversation between humans and machines, with an eye toward the distinctive ways turn exchange may be managed across cultures. Even less well understood is how human-machine turn exchange is accomplished in interactional contexts where multiple potential communicative modalities are at play. Much more attention has been paid to how systems use sound (Brewster, 1998; Rinott, 2008), recognize and produce speech that invokes human emotion (Busso, et al., 2004; Cahn, 1990; Mor, 2014; Oakley et al., 2000), and more recently, operate in a multi-modal capacity (see Dumas et al. 2009; or Wechsung, 2014, for a review).

This chapter seeks to address this gap through an analysis of trouble in turn exchange in human-computer multimodal interaction in an in-car infotainment system in two cultural contexts. The first in a suburban region of New England, the second in an urban center in China. One central question we examine is how, in these interactions, where participant expectations of their interlocutor's competence as a model interactant may not hold, does repair get done, and what does the trouble and its repair tell us about the culturally distinctive ways to do it "right"?

In what follows, we review the concepts and theoretical framework employed in the analysis and research design that produced the dataset we analyze here, though the theory and methodology as adapted for the study of in-car communication is more fully detailed in Chapter 2.

Next, we analyze a number of instances to determine the source of trouble experienced by many participants in interaction with the system, and the variety of methods users employed for accomplishing their goals despite this trouble. Based on this analysis, we highlight cultural norms and premises governing interaction in this communication situation (Hymes, 1972). We conclude with a discussion of the implications for multimodal system design.

5.2 Theoretical Framework and Related Literature

In the tradition of Conversation Analysis (see Heritage, 2010, for a summary of principles), it has long been accepted that conversation occurs in a sequential fashion, organized through the managed exchange of turns at talk, though debate exists over seemingly contradictory cases (Reisman, 1974; Sidnell 2001). Generally, this organization is taken to be fundamental to meaning-making in interaction. Schegloff (2000) captures this stance in the following:

> The orderly distribution of opportunities to participate in social interaction is one of the most fundamental preconditions for viable social organization. ... One feature that underlies the orderly distribution of opportunities to participate in conversation, and of virtually all forms of talk-in-interaction that have been subjected to disciplined empirical investigation, is a turn-taking organization. The absence of such an organization would subvert the possibility of stable trajectories of action and responsive action through which goal-oriented projects can be launched and pursued through talk in interaction... (p. 1)

The preponderance of articles on the topic of turn-based organization preclude a thorough review here, but for the seminal work of Sacks, Schegloff and Jefferson (1974). Therein, the authors propose a Turn Constructional Unit (TCU) and a Turn Allocation Component. The idea of the TCU suggests that interactants' turns are constructed in such a way as to make the kind of turn it is, the action the turn seeks to accomplish, available to fellow interlocutors such that the projection of the coming end of the turn can be anticipated. That such a function exists in conversation is evidenced by the ability of interlocutors to cut-in before a turn is fully completed, having projected what the completed utterance may likely have contained. The Turn Allocation Component suggests that interlocutors actively manage the exchange and allocation of turns at talk through a variety of practices that are designed to select a next speaker, or self-select as next speaker, and signal when a speaker's turn is completed, or about to be, through a transition-relevance place (TRP). Schegloff (1992), conceptualizes TRPs as "discrete places in the developing course of a speaker's talk (...) at which ending the turn or continuing it, transfer of the turn or its retention become relevant" (p. 116).

The sequential organization of speaking turns also provides the foundation for the interpretation of the meaning of talk in interaction. For instance, returning to our hypothetical conversation between friends in the introduction, if "what concert do you want to see?" is the first turn in a conversation, its meaning may be heard as an invitation to a concert. However, following a prior turn where a friend asks "do you want to see a movie?" that same speech may now be interpreted as indication of a mishearing of the prior turn, or a rejection of the invitation to a movie and a counter-invitation to a concert. Since meaning is reliant on the position of an utterance in relation to surrounding utterances, the timing of conversational turns becomes significant in managing the mutual intelligibility of the interaction, without which the interaction cannot continue without repair. How humans and machines in interaction manage this exchange of turns, and the careful timing required to do so in order to assure mutual intelligibility is of primary concern to this work.

One way interactants can signal that a turn is complete is the use of pause (Maynard, 1989). An interlocutor may use a pause at the end of an utterance to invite another speaker's participation, but they may also simply be pausing for breath, to develop their next utterance, because they were distracted, etc. The trouble, then, is determining whether a gap in speech is an invitation to exchange speaking turns, or merely a period of silence where the speaker intends to maintain the floor. Among other strategies, like audible in-breaths, or other disfluencies (Corley & Stewart, 2008) such as "umm" or "annnd," culturally competent interlocutors come to know the length of time an interactant might pause to signal a TRP and actively monitor for these in conversation, though syntax (Sacks, 1974), prosody (Couper-Kuhlen & Selting, 1996), pointing (Mondada, 2007) and other pragmatic information (Ford & Thompson, 1996) are also potential cues of the coming completion of a turn.

Indeed, in instances of intercultural interaction differences in the use of silence can often create trouble as speakers evaluate the meaning of silence from a cultural vantage. Carbaugh (2005), documents a moment of such trouble in an introductory meeting with a future colleague in Finland, Jussi Virtanen. To Carbaugh's surprise, Virtanen would respond to each of Carbaugh's turns at talk with a 10-20 second pause. From the vantage of an American English speaker from the Northeast, these pauses were exceptionally long and, for Carbaugh, signaled something untoward in the interaction. As he later discovered, the pauses were the result of a confluence of factors including a Finnish customary practice of long (from the American view) pauses after sentences, Virtanen's personal use of longer pauses (from the Finnish view), Virtanen's careful use of English as a second-language, and finally the use of long pauses as a means of signifying the respect one has for the occasion and its significance. Here, then, inter-turn pause is both the result of situational factors, but also a motivated use of cultural means for communicating respect and appreciation.

Scollon and Scollon (1981) note trouble with culturally distinctive pause lengths in conversation in their study of interaction between Athabaskan-English speakers and U.S. American-English speakers. The authors find that in conversation,

Athabaskans are often overrun by English speakers because of a preference for pausing between the exchange of speaking turns for about a half second longer than typical for U.S. English speakers. This means that English speakers often set the topic of conversation, and then proceed to dominate (from the Athabaskan view) the remainder of the conversation as Athabaskans monitor for TRP's at the end of the English speaker's turn, only to find that the English speaker has started speaking again before they had a chance to take their turn. This leads to negative evaluations of the conversation from both the U.S. English and Athabaskan-English speaker's view, based primarily on cultural variation in inter-turn pause length and the meaning of pauses that last relatively longer or shorter.

Because turn-taking practices serve as the foundation for mutual intelligibility in conversation, and these practices are subject to cultural variation (Tannen, 2012), understanding the cultural norms and premises governing the exchange of turns at talk is essential. Attending to issues of variation in norms for turn taking can help illuminate trouble in human computer interaction in the same way such an analysis can illuminate trouble between members of different cultures. As a result, in the analysis that follows, we make use of concepts from Conversation Analysis, reviewed above, as well as a framework for analyzing the culturally distinctive ways talk is patterned in interaction – Cultural Discourse Analysis (Berry 2009; Carbaugh 1988, 2007, 2012; Scollo 2011).

Cultural Discourse Analysis (CuDA) is a development of the Ethnography of Communication (Hymes, 1962, 1972, 1974), that seeks to describe, interpret, compare and critique culturally patterned communication practices. CuDA invites us to investigate communication acts, events, and situations for "radiants of meaning" found in messages about personhood, social relations, emotion, place, and communication itself. Presumed and enacted in these messages are "cultural premises" which serve as a resource for the interpretation and production of meaning in interaction. Cultural norms may also be identified, which are implicit or explicit rules that govern the moral domain of social action. In our analysis, we employ these concepts in understanding the distinctive premises and norms that both system designers, and users of systems, may employ as they seek to accomplish their respective goals for the interaction, as well as the ways they may sometimes be misaligned, the consequences of that misalignment, and what users do to get back on track.

5.3 Methodology

As discussed in previous work (Carbaugh, et al. 2012, 2013; Molina-Markham et al. 2014, 2015), and reviewed in chapter 2, data for the analyses below were collected from the driving sessions of 26 (14 female, 12 male) participants during the study of an in-car infotainment system conducted in Western Massachusetts. During the driving session, participants would use their own car, which had been outfitted with a prototype infotainment system that they would interact with through a dashboard mounted tablet computer. Participants were asked to drive on mostly rural roads of their choosing for one and a half to two hours on average. During this time participants were invited to make use of a variety of the voice capabilities of the system as they would in the normal operation of their own vehicle, were they to have such a system. Touch interaction with the system was permitted for starting or ending interactions through use of the microphone button, which started a voice command event, or the "end" button, which could be used to close an ongoing action.

Being interested in the sequential organization of multimodal interaction in this context, the way users managed turn exchange with the system, and any potential misalignment between the system's behavior and users' cultural norms and premises operating in this communication situation, we identified interactional sequences where the system's talk overlapped with the user as a sign of potential turn exchange trouble. We found that when users engage in a task-switching event (asking the system to do a task that is not part of the current task the system is performing) a disproportionate amount of overlapping talk occurred relative to other interactional sequences, like directing the system to perform a new task. Thus, the data for this analysis are taken from overlapping talk that occurred during user attempts to switch tasks.

Not all users made use of the task switching functionality of the system, which produced a corpus of 7 participants that did, each of whom experienced some degree of overlapping talk with the system on their first use of the task-switch event, making this a regular and robust phenomena for investigation.

5.4 Prompt Timing and Misalignment – A Formula for Interruptions

When the system was engaged in the ongoing performance of a given task (playing the radio, making a phone call) and the user pressed the microphone button to initiate a voice command, the system was programmed to respond by providing an audible "ding" sound to confirm that the system received the users request to initiate a voice command and was now in an "on" state. The system would then play a task-relevant prompt. For instance, while in the radio task, if the microphone button is pressed, the system would ding and then ask "which station or channel do you want to hear?" It was then the user's turn to talk. However, the sequence never went according to design in the first instance. Below is an example of a typical way this interaction occurred.

Instance 1

Participant 11 - 51:44

1   P: (participant touches microphone button)
2   S: (audible ding)
3   (0.6)
4   P: phone ca[ll
5   S: [which station or channel do you want to hear?
6   P: phone call

In this instance, the participant presses the microphone button while listening to an FM radio station. The system responds to the participant's touch with an audible ding. There is then 0.6 seconds of silence before the participant begins her directive to the system -- "phone call." During this directive the system overlaps her talk with its own prompt "which station or channel do you want to hear? The participant responds by restating her directive from the prior turn, "phone call," on line 6, which can be understood as a corrective action since the system's turn was not responsive to her command "phone call". In this way, the system issues an utterance that is not sequentially relevant to the participant's prior turn. In this instance, the user opts to treat the system's violation as the result of a mishearing and reissues her command.

Because the system is designed to take the first turn, it is not listening for the participant's voice between the audible ding and its first turn, the verbal prompt. Therefore, it cannot move to cut-off its turn in recognition of the participant-issued directive as a human interlocutor might (Schegloff, 2000). Because the participant waits only .6 seconds before beginning their turn, they issue a command that is not heard by the system, and which causes the system's turn (its first turn from the system's view) to be badly non-responsive to the sequential position of the interaction at that juncture.

Now that we understand what the system thinks is happening we might ask, why does the participant take their turn at .6 seconds and not wait for a verbal prompt from the system, which in this case occurs after 1.5 seconds? One possible explanation is misalignment between the system and participant on the meaning of the audible ding on line 2. The participant may understand this ding in a number of ways. We first suggest that the participant might understand the ding as a summons response, borrowing from the interactional form of the telephone conversation.

In 1964, Sacks (Jefferson & Schegloff, 1995) pioneered studies of telephone call interactions and concluded that the ringing telephone functions as a summons and the answering of the phone with "hello" functions as a response to the summons (Sacks, 1974), ostensibly two turns at talk have been exchanged, the "ring" and "hello." The next turn, then, wherein the topic of the conversation is set, belongs to the actor who did the summoning, the caller. In the car, a participant issues this summons through touching the microphone button, and the system responds with an audible ding. If this interaction were following the routine form of the telephone call then the participant would take the next turn and set the topic. Instead, here, the system attempts to set the topic by asking what station the participant wants to hear. It is possible, then, that participants are modeling interaction with the system after the routine form of the telephone call, wherein the participant takes the first turn at talk, and that this understanding informs their move to initiate their directive before the system's verbal prompt, since they do not expect the system to be taking a turn in this position. We suggest that this capacity of users to borrow meaning and form from routine interactional practices in one context in order to help make sense of novel contexts be understood as a kind of *cross-pollination* of user experience and sense-making.

Also at work are cultural norms for the amount of time that passes in a gap between turns before that gap signifies a Transition Relevance Place (TRP) where a conversational turn may be understood to be over or relinquished. If a cultural norm exists for the participant that turns are exchanged after roughly .6 seconds of silence, then the system will routinely take too long to take its first turn as users proceed to interpret the system's silence as yielding the speaking floor. As we see in the following instances this appears to be the case.

Instance 2

Participant 8 - 28:06

1 P: (Participant touches microphone button)
2 S: (audible ding)
3   (1.0)
4 P: next
5 S: what artist would you like?
6 P: next

In this instance, like the last, the user is in a task, in this case listening to their downloaded music library, when she decides to touch the microphone button. After doing so the system dings, and after a 1 second gap she issues her directive "next." The system's next turn, which sequentially would be heard as a reply to the participant's directive, asks the participant what artist she would like to hear. This verbal prompt is of course not responsive to the participant's directive, and so the participant restates it on line 6, again treating the system's turn as the result of a mishearing in need of correction. The amount of time it takes the system to respond to the microphone button press was measured at 2 seconds; this is the longest the system takes to issue a prompt. This allows the participant to wait 1 second and then issue her directive in the remaining 1 second before the system plays its prompt. Because of this the user does not experience overlapping talk with the system, but is perhaps presented with an even more confusing response, since the overlap itself can function to let the participant know that something is wrong. Without the benefit of the overlap, the user is left to wonder what the system's turn means and what should be done next?

This 1 second pause, like the .6 second pause in the prior instance, is long enough to indicate to the participant that the system has yielded the floor, and it is now her turn. This is not the case however, given the system's design, and the system proceeds to issue what it takes to be its first turn, leaving the participant to conclude that the system has either not heard, or misheard her command. This may negatively impact participant perception of the competence of the system as a voice interaction partner.

Across these instances, participants encountered the same interactional trouble, attempting to initiate a directive to the system that the system responds to with a sequentially non-responsive verbal prompt, and/or in most cases, the participant's talk is overlapped by the system's forcing the participant to compete for the floor or abandon the turn.

Participants in the larger corpus from which these instances are taken varied in the amount of time they waited before speaking after the audible ding from the system from 0.6 seconds to 1.5 seconds. Since the average time the system takes to generate a prompt following the audible ding is 1.7 seconds, this means those participants who wait around 1.5 seconds to give the system a directive will almost certainly be interrupted by the system, while those who begin a directive immediately following the ding may be able to complete their directive only to be met with a question that seems irrelevant to the directive they have just issued. A seemingly simple fix for this trouble is an anti-overlap feature to assure that if the system hears the user talking, it hold its turn until the system can decide what next action to take that would be relevant to the user's speech. However, listening for user speech all the time when it has no reasonable expectation that the user is about to speak means lots of mistaken "hearings" on the system's part that could lead to even more trouble.

The patterning of the interaction above is suggestive of a norm for the management of turn exchange in conversational positions where turn allocation is ambiguous. This norm treats pauses of longer than .6 seconds, and no longer than 1.5 seconds, to be indicative of the passing of a turn. The system's routine pause length of 1.3-2 seconds, then creates a misalignment in the turn-taking management of interaction between the participant and system. As a result, participants are forced to abandon their turn, or competitively produce a turn in overlap with the system. Participants must further make sense of the system's turn, which given its late positioning in the interaction relative to the position it was designed to inhabit, appears non-responsive to the participant's directive. In the instances above, participants treated the system's turn as a mishearing in need of correction through repetition of the initial directive, though this was not the only way participants managed to negotiate the difficulty of this misalignment. After having encountered this misalignment some number of times participants would generally adjust their interaction with the system in one of 4 ways, which we review in the following section.

5.5 Interactional Adaptation

After a participant experienced the system overlapping their directive, and/or re-sponding to their directive in non-responsive ways, they appeared to learn, at differ-ent rates, that the system will be taking a turn after the audible ding in task-switch events, and that this turn will take place after some notable pause. Given their appar-ent noticing that this is the case, users proceeded in future interactions with the sys-tem in one of 4 ways.

5.5.1 User Adaptation: Competitive Production

One way participants adapted to the system was to sustain a "competitive produc-tion" (Schegloff, 2000). In the instance below this is accomplished through the extra-ordinary elongation of the vowel sound in "too," sustained until after the system's turn had completed.

Instance 3
Participant 10 - 1:24:33
1 P: (Participant touches microphone button)
2 S: (audible ding)
3   (0.8)
4 P: change st[ation too::owuh (1.2)
5 S: [what radio station do you want to hear?
6 P: ninety seven point three
7   (4.5)
8 S: could you repeat that please?

In this case, the participant refuses to abandon their turn to the system and makes a bid to hold the floor through the elongation of the vowel sound in "too" on line 4. A fellow human interactant would then be forced to choose to continue their own turn in a sustained overlap, or yield the floor to the other speaker. Because the system is not listening when it's playing its own prompt, it is incapable of knowing that the participant is speaking and therefore incapable of deciding to abandon its turn. This means the system, in instances of competitive production, will always sustain overlap until its turn is complete. Somewhat ironically, however, the system can never "win" since human interactants engaged in competitive production can project the incipi-ent end of a turn shape and adjust their strategy for elongating their turn to assure it lasts longer than the system's. This is the case in the instance above.

Despite the participant "winning" the competitive production, the directive to tune to 97.3 cannot be understood by the system because it was not listening to the participant's utterance during the overlap. Even if the system had been listening it would not be able to understand a directive including the extraordinary stretched vowel seen here. A bitter-sweet victory. The implications for the outcome of this competition among human interactants would likely include messages about the status of the relationship between interactants. As Tannen (1993) points out, however, the meaning of this overlap to participants is not set *a priori* as Schegloff's (2000) use of the term "competitive production" might suggest. Overlap may also be understood by interactants as a move to solidarity, though it does appear in this instance that the participant intends to outlast the system. Regardless, the possible interpretations of the meaning of the overlap to participants, one thing is certain, the system will take no implications about social relations from the interaction, though the participant may.

This is one way participants have borrowed interactional strategies such as competitive production from human interaction for use in dealing with an invasive conversational partner (the system) but where the social implications of that strategy may not carry over. As a result, this can be seen as an additional instance of the phenomena of *cross-pollination* as users draw on familiar interactional dynamics to navigate this novel interactional trouble.

5.5.2 User Adaptation: Defer to System

The second way users adapted to the system's overlap was to wait out the long pause for the prompt and then speak. In this strategy, participants allow for a longer pause than they generally had in the past, giving the system the opportunity to play its prompt. The participant would then give their directive, which was necessarily shaped as a corrective, since quite often the directive they gave the system was not related to the task-specific prompt the system played.

Instance 4
Participant 8 – 31:08
1   P: (Participant touches microphone button)
2   S: (audible ding)
3   (1.3)
4   S: What artist would you like?
5   P: FM Radio
6   S: Just a second

Here the user touches the microphone button and the system dings in reply, 1.3 seconds pass, and the system asks the participant what artist she would like to hear. The participant, apparently not wanting to hear an artist replies "FM Radio". Structurally, the exchange of turns has gone smoothly here (no overlap) though there are two things to note. First, this participant experienced trouble with a turn exchange performing a task switch 4 minutes prior (Instance 2), experiencing the overlap phenomenon common among all users. As a result, we suggest that her decision to not take a turn during the 1.3 seconds gap after the system's "ding" is an adaptation to the overlap trouble from her prior task-switch experience. In this case, the participant adapts to the system's norm for a 1.3-2 second pause before its first turn. In so doing, the participant has moved through a process to identify the trouble (the system intends to take a turn at talk after the ding, and has a relatively long pause before it does so), develop a possible solution (wait until the system speaks), and implement that solution, though it contradicts her and other participants' routine norm for managing turn exchange (a 0.6-1.5 second pause).

Despite having adjusted the timing of her initial turn to accommodate the system, she is still placed in the position of having to respond to the system's prompt with a move to reject the system's offer. Since the system opts to take a guess at what the participant might want, asking "what artist would you like?" the system constructs its turn as to prefer a response that chooses an artist. Any response from the participant that is not the name of an artist is thereby shaped as a dispreferred response (Heritage, 1983; Levinson, 1983; Pomerantz, 1984). This restricts the available next actions for the participant to either a response to the question that selects an artist, or an outright rejection of the system's offer to play an artist, which people would generally rather not have to do.

5.5.3 User Adaptation: Abandon Voice Task Switching

Another option participants developed for dealing with the overlap involved an abandonment of the use of the task switch capacity. In this sequence, participants opted to switch tasks by first touching the "End" button to stop the current task (playing the radio), and then pressed the microphone button to initiate a new voice command. When the microphone button is pressed outside of an ongoing task, like when the user is on the system's home screen, the system plays an audible ding and then waits for the participant's command. Possible overlap is then avoided by selecting an interactional path that does not include the system taking a spoken turn. One benefit, then, of multimodal systems is the ability of the user to adapt to verbal interactional trouble by employing alternate modes that avoid the trouble.

Instance 5
Participant 9 - 53:05
1   P: (Participant touches End Radio button)
2   (2.0)
3   S: (Radio stops playing, screen shifts to home)
4   (.5)
5   P: (hand begins move toward radio)

6   P: (.7)
7   P: (finger touches mic button)
8   P: gimme dubbelyu=efem (.) ehhhn give me doubelyu::::::
9   whatsitcalled effseear

Here the user begins by ending an ongoing task, the playing of the radio by touching the End Radio button. It takes the system 2 seconds to comply with the user's directive to stop playing the radio and return to the home screen on the display. Within .5 seconds the user begins to move his hand back toward the radio and makes contact with the microphone button .7 seconds later. He then gives the radio a directive to play WFCR, all in less time than it took the system to comply with his initial directive to end the radio.

It is likely, then, that the user intended to change radio stations when he hit the end radio button, but why not just press the microphone button and tell the system to change stations, making use of the system's task-switch function? The answer we propose that best accounts for the participant's actions here is that in prior interactions the user had difficulty with the turn exchange, particularly, during a task switch event 40 minutes prior, whereafter he ceased to use the task-switch functionality, opting instead to explicitly end all ongoing tasks through touch before initiating a voice interaction to issue a new command. Abandoning the line of interaction that produced the turn-exchange difficulty is then one method a user developed for accomplishing the desired task, despite trouble with the timing of turn exchanges in the task-switch event. Doing so is likely not the ideal case however, as the task switch function allows users to achieve their goal in as little as one button press and one voice command, while the strategy adopted by this participant will require a minimum of two button presses and a voice command. This is not ideal from the perspective of system designers either, since the minimization of the use of touch while driving is preferred for safety reasons.

5.5.4 User Adaptation: Persistent Trouble

Not all users did develop new methods for dealing with the overlap trouble. One participant continued to repeat the pattern observed in instances 1 and 2 (issue command, system overlaps, reissue command) 6 times repeatedly, one after the other, over the course of her drive with the first occurrence at minute 20 and the last 42 minutes later. During this time the participant never adjusted the pattern of her interaction, continually experiencing overlap with the system each time she performed a task switch. We have included one of these instances here for illustration, though the patterning is identical to instances 1 and 2 reviewed above. The following is taken from the fourth recurrence of this pattern with this participant.

Instance 6
Participant 11 – 54:19
1   P: (Participant touches microphone button)
2   S: (audible ding)
3   P: phone call
4   S: which station or channel do you want to hear
5   P: phone call↑
6   S: okay (1.0) who would you like to call

That this participant persists across a number of instances to issue a directive prior to the system's turn is likely the result of the participant never identifying that the system intends to take the first turn at verbal interaction and as a result is not listening as it prepares its turn. Instead, the participant through the repetition of her initial directive, treats the system's prompt as a mishearing of her initial directive in need of repetition. If the participant had identified that the system was not listening, persisting with issuing the command before the system's prompt would serve no purpose and would likely have discontinued. This is suggestive that not all participants are equally adept at identifying the source of trouble with the system's behavior. This instance further highlights the trouble of poorly crystalized norms surrounding the meaning and turn status of the audible "ding" in multimodal interaction.

5.6 Norms and Premises

The analysis above suggests some normative ways that participants approach interacting with the system, as well as certain premises that inform this use. In the instances presented above, each participant experienced either an overlap of talk with the system and/or a seemingly non-responsive reply to their directive. The regularity with which this phenomenon occurred throughout the corpus suggests that the amount of time these participants understand as evidence of, or opportunity for, a turn exchange, in contexts where next speaker is ambiguous, is less than the 1.7 second average time the system takes to produce its verbal prompt. We believe this amount of time to be a cultural norm for managing the interactional exchange of conversational turns when the next speaker is ambiguous. The system, then, is engaged in a kind of norm violation when it produces its overlapping prompt that carries the interactional force of an interruption, violating the moral order of turn-taking and politeness that is generally expected between human interactants in social interaction. This norm can be more explicitly formulated as: *In contexts where next speaker is ambiguous, if an interactant wishes to take a turn, they should do so between .6 and 1.3 seconds after the prior action.*

The cultural nature of this norm is evident in the results of cross-cultural research that investigated Chinese participants use of the system in Shanghai and Beijing.

While the results of participants overlapping speech with the system in China is discussed more fully in chapter 6, we note here that the average time Chinese users waited after the system's initial ding ranged from 0.5 to 0.9 seconds. This suggests that Chinese participants are likely operating under a cultural norm for pause length that indicates a TRP that is significantly shorter than in the U.S. The result of this shorter pause between turn exchanges in the Chinese interactional data led to a relatively greater number of participants overlapping with the system, even when the system was producing its verbal prompt at its fastest capacity. This in turn led to Chinese participants evaluating the system as generally "too slow."

However, misalignment between the system and participants was not restricted solely to the normative timing of the exchange of speaking turns, but also in the meaning of particular actions within a communication event, as in the case of the audible ding. Whereas the system was designed with the audible ding's intended meaning being an alert of the system's status, akin to announcing the system is on, participants treated the audible ding as a summons response akin to the organization of telephone calls. Depending on which meaning of the audible ding one employed, a different next speaker would be appropriate. From the participants' vantage the audible ding occupied the space of an interactional turn, and therefore the system was understood to have passed the turn back to the participant for their first spoken turn.

In all but one case, participants chose to adjust their interactional strategies for accomplishing the task they sought, with one user persisting in the original pattern of overlap, likely doing so as the result of failing to identify that the system was not listening in the gap after the audible ding. This means that all participants who became aware of the source of the trouble opted to make adjustments in order to accomplish the task.

It is not automatically the case that this should be so. In human interaction, an interlocutor behaving in the way the system does would likely be called to account for their repeated interruptions for no apparent conversational purpose. However, the participants in the instances collected above never call the system to account for its behavior, nor exhibit any animus toward the system for what might be cause for an argument with a human interactant. The system, in effect, gets a pass. This is not to say participants will, or do, find interacting with a system under these conditions pleasing, only that the system itself appears not to be held responsible for its behavior in this context. This is likely because participants understand that the system lacks the fully-fledged capabilities of a culturally competent human interactant, and therefore cannot be held responsible for these sorts of issues, but likely not trusted either.

A premise of and for communication can then be identified in the participants' interaction that *those who are not fully competent interactional partners cannot be held responsible for certain interactional blunders.* An accompanying premise of personhood can then also be formulated as *voice interactive machines are not fully competent interactional partners.* And finally, an additional norm can be identified for proper behavior given the above premises, *since machines are not fully competent interactional partners, human users ought adjust to the system in order to accomplish their goals.* These premises likely inform the level of tolerance users have for interacting with systems that routinely violate human interactional norms, without which interaction with systems at this level of capability would not be possible. This does not mean that the above premises are universal or automatic.

In data collected from interactions with the system in mainland China participants demonstrated a regular willingness to overlap the system's verbal prompt, repeatedly issuing their own directives while the system was speaking. In the instances reviewed above, all but one participant, after encountering one of these unintentional overlaps, developed an alternative strategy to avoid this overlap and accomplish their desired action. However, in the Chinese participant data these adaptations are less frequent with the majority of users persisting in their overlaps with the system. While the overall greater number of overlaps in the Chinese data are likely due to the apparent Chinese cultural preferences for shorter inter-turn pauses, that Chinese users continually overlap the system in this manner suggests that something more than misalignment on the timing of turn exchange is happening here.

While U.S. American participants in this study orient to the system as an interactional partner with somewhat diminished capacity and competency, Chinese participants appear to orient to the system as a subordinate, akin to service staff who should be better able to adjust to them in satisfying their desires. As a subordinate, the system should yield discursive power to the participant in terms of when to talk (pacing), what to talk about (dialog topics), and style of interaction (efficient or elaborate). This orientation to the system as a kind of subordinate staff may inform the perceived legitimacy of repeated overlaps with the system, and a greater demand that the system abandon its turn and yield the floor to its superior. The U.S. participants' orientation to the system as a limited capacity machine who cannot be held to the same expectations as human interactants, may then allow for a higher level of acceptance of failed adaptations on the part of the system necessitating a higher number of adaptations on the part of the user. Informing the observed dynamic in these cross-cultural contexts, then, are likely a differing set of cultural premises about the nature of the system being interacted with, and the rights and obligations of human interactions engaged with it. These distinctive premises are identified in the comparative table below.

Tab. 1: Comparison of US American and Chinese mainland premises and norms

U.S. American premises and norms for HMI	Chinese premises and norms for HMI
US American premise for personhood: voice interactive machines are not fully competent interactional partners.	**Chinese premise for personhood:** voice interactive machines are subordinates who attend to the human interactant's needs.

U.S. American premises and norms for HMI	Chinese premises and norms for HMI
US American premise of relations: since machines are not fully competent interactional partners, human users ought to adjust to the system in order to accomplish their goals	**Chinese premise of relations:** since machines are there to service human interactants, they need to adjust to human interactants' communication styles in order to meet their needs.
US American premise for communication: those who are not fully competent interactional partners cannot be held responsible for certain interactional blunders.	**Chinese premise for communication:** voice interactive machines need to be able to efficiently respond to commands and requests and do so in a pleasant, service-oriented manner.
US American norm 1: In contexts where next speaker is ambiguous, if an interactant wishes to take a turn, they should do so between 0.6 and 1.3 seconds after the prior action.	**Chinese norm 1:** In contexts where next speaker is ambiguous, if an interactant wishes to take a turn, they should do so between 0.5 and 0.9 seconds after the prior action.

One potential consequence of these differing cultural premises may be a more severe evaluation of the system's competence among Chinese users, leading to higher levels of user dissatisfaction than among U.S. participants who generally appear more accepting of the system's inability to relinquish the floor in instances of overlapping speech.

5.7 Implications for Design

The analysis above can be used to make particular recommendations for the improvement of multimodal interactive systems in the future. First, the task-relevant prompt is problematic as many participants noted in interviews that it was unnecessary, inappropriate, or too long. Some participants suggested that no prompt was needed at this stage of the interaction at all, citing that when someone presses the microphone button while in a task they likely have something they would like to do in mind, and have pressed the microphone button in order to give the system that command. As a result, the system need not offer any prompt, but rather just listen for the participant's command.

Second, system designers need better understand the role of non-speech sounds in multimodal interaction and their turn taking relation to other modalities such as touch and speech. In the design of this system, the audible ding is treated as if it occupies no conversational position -- it takes no turn. This is clearly not how participants in the above interactions understand the ding. A turn-based analysis of the interaction suggests that the ding does function as a turn-at-talk, with the first turn being the participant's touch of the microphone button, the second turn being the system's reply to the touch through audible ding, and the third turn then passing back to the user for first topic. However, because the system design does not account for the audible ding as an interactional turn, it presumes users will wait for the system to respond to the microphone press with a verbal prompt. This appears to not be the case as users hear the audible ding as the response to the microphone press and proceed to take their turn. Some research on the role of non-speech sounds in human-computer interaction is already underway (Brewster, 1997; Brewster, 2002; Hereford & Winn, 1994), but does not incorporate an analysis of the sequential position of this mode in the organization of interaction.

Some participants did report a desire to have a system prompt after the microphone button was pressed as a sign that the system is "listening," but thought the prompt that was offered was simply too long for regular use. Participants suggested alternate prompts including "yes?" or "what would you like?" which are likely better alternatives as they are task independent and do not require users to reject the system's wrong guess, which as indicated above is a dispreferred action in conversation. Any verbal action taken after the ding would also need to be performed quite quickly so as to avoid the potential for overlap, particularly among cultural members with shorter inter-turn pause lengths.

Ultimately, however, what holds multimodal interactive systems back the most in the instances analyzed above is the system's inability to listen for user speech and act accordingly. Cases of overlap in human interaction are resolved in a variety of ways (Schegloff, 2000) but all require monitoring of the ongoing turn by both interactants. In order to properly model human interaction, the system must be able to listen to users' ongoing turns and adapt, as we do with them. Research on the broader phenomena of overlapping speech in HCI, sometimes referred to as "barge-in," is also underway, examining the frequency and context of "barge-in" cross-culturally, as we do here in chapter 6.

We further advocate attention be paid to the cultural nature of the management of turn-exchange both in the amount of time interlocutors normatively wait as indication of a TRP, but also in the practices employed managing turn exchange, and the strategies adopted to accomplish interactants' goals. The analysis above suggests two cultural norms and two premises of and for communication and personhood that may vary culturally and influence the way users interact with these kinds of systems, particularly surrounding the resolution of trouble and the meaning of that trouble.

References

Berry, M. 2009. The social and cultural realization of diversity: An interview with Donal Carbaugh. Language and Intercultural Communication, 9:230–241.doi:10.1080/1470847090 3203058

Bossemeyer, R. W. and Schwab, E. C. 1990. Automated alternate billing services at Ameritech. Journal of the American Voice I/O Society 7:47-53.

Brewster, S. A. 1997. Using non-speech sound to overcome information overload. Displays 12(3-4):179-189.

Brewster, S.A. 1998. Using non-speech sounds to provide navigation cues. ACM Transactions on Computer-Human Interaction 5(2):224-259.

Brewster, S. A. 2002. Nonspeech auditory output. In: (A. Sears, and J. Jacko ,eds) The Human-Computer Interaction Handbook: Funamentals, Evolving Tehcnologies. CRC Press, pp. 221-237.

Brown, P., Levinson, S. C. 1987. Politeness: Some universals in language usage. Cambridge University Press. Cambridge.

Busso, C., Deng, Z., Yildirim, S., Bulut, M., Lee, C.M., Kazemzadeh, A., Lee, S., Neumann, U., and Narayanan, S. 2004. Analysis of emotion recognition using facial expressions, speech and multimodal information. Proceedings of the 6th International Conference on Multimodal Interfaces, State College, pp. 205-211.

Cahn, J. E., The Generation of Affect in Synthesized Speech. Journal of the American Voice I/O Society 8:1-19.

Carbaugh, D. 2007. Cultural discourse analysis: Communication practices and intercultural encounters. Journal of Intercultural Communication Research 36:167–182. doi:10.1080/17475750701737090

Carbaugh, D. 2012. A communication theory of culture. In: (A. Kurylo, ed) Inter/Cultural Communication: Representation and Construction of Culture, Sage. Thousand Oaks, pp. 69-87.

Carbaugh, D., Molina-Markham, E., van Over, B., and U. Winter. 2012. Using communication research for cultural variability in human factor design. In: (N. Stanton, eds) Advances in human aspects of road and rail transportation. CRC Press. Boca Raton, (FL), pp. 176–185.

Carbaugh, D. and Poutiainen, S. 2005. Silence, and third-party introductions: An American and Finnish dialogue. In: (D. Carbaugh, ed) Cultures in conversation. Lawrence Erlbaum Associates. Mahwah, pp. 27-38.

Carbaugh, D., Winter, U., van Over, B., Molina-Markham, E. and S. Lie. 2013. Cultural analyses of in-car communication. Journal of Applied Communication Research 41(2):195-201.

Corley, M., and Stewart, O. W. 2008. Hesitation Disfluencies in Spontaneous Speech: The Meaning of um. Language and Linguistics Compass 2(4): 589–602.

Couper-Kuhlen, E., and Selting, M. 1996.Towards an interactional perspective on prosody and a prosodic perspective on interaction. In: (E. Couper-Kuhlen and M. Selting, eds) Prosody in Conversation. Cambridge University Press. Cambridge, pp. 11-56.

Dumas, B., Lalanne, D., and Oviatt, S. 2009. Multimodal interface: A survey of principles, models and frameworks. In: (D. Lalanne and J. Kohlas, eds) Human machine interaction: Lecture notes in computer science. LNCS-Springer-Verlag. Germany, pp. 3-26.

Ford, C. E ., and Thompson, S. A. 1996. Interactional units in conversation: Syntactic, intonational, and pragmatic resources for the management of turns. In: (E. A. Schegloff and S. A. Thompson, ed) Interaction and grammar. Cambridge University Press. Cambridge, pp. 135-184.

Grice, H.P. 1975. Logic and Conversation. In: (P. Cole and J.L. Morgan, eds) Syntax and Semantics 3:41-58.

Hereford, J. and Winn, W. 1994. Non-speech sound in human-computer interaction: A review and design guidelines. Journal of Education Computing Research 11(3):211-233.

Heritage, J. 1983. Garfinkel and Ethnomethodology. Polity. Oxford

Heritage, J. 2010. Conversation Analysis: Practices and Methods. In: (D. Silverman, ed) Qualitative Research (3rd Edition). Sage. London, pp. 208-230.

Hymes, D. 1972. Models for the interaction of language and social life. In: (J. J. Gumperz and D. Hymes, eds) Directions in sociolinguistics: The ethnography of communication. Blackwell. New York, pp.35–71.

Hymes, D. 1974. Foundations in Sociolinguistics: An Ethnographic Approach. University of Pennsylvania Press. Philadelphia (PA)

Jefferson, G. and Schegloff, E. 1995. Lectures on Conversation. Willey Blackwell.

Levinson, S. 1983. Conversational Structure. Cambridge University Press. Cambridge.

Maynard. S. K. 1989. Japanese conversation: Self contextualization through structure and interactional management. Ablex. Norwood (NJ)

Molina-Markham, E., van Over, B., Lie, S. and D. Carbaugh. 2015. "OK, talk to you later": Practices of ending and switching tasks in interactions with an in-car voice enabled interface. In: (T. Milburn, ed) Communicating User Experience: Applying Local Strategies Research to Digital Media Design. Lexington Books. London. pp. 7-25.

Molina-Markham, E., van Over, B., Lie, S. and D. Carbaugh. 2016. "You can do it baby": Cultural norms of directive sequences with an in-car speech system. Communication Quarterly 64(3):324-347.

Mondada, L. 2007. Multimodal resources for turn-taking: pointing and the emergence of possible next speakers. Discourse Studies 9(2):194-225.

Mor, Y. 2014. The future of human-machine interaction: It's not what you say, it's how you say it. Wired. Retrieved from: http://www.wired.com/2014/02/future-human-machine-interaction-say-say/. Accessed: 2/21/14.

Oakley, I., Brewster, S.A. and Gray, P.D. 2000. Communicating with feeling. In: Proceedings of the First Workshop on Haptic Human-Computer Interaction, pp. 17-21.

Pomerantz, A. 1984. Agreeing and disagreeing with assessments: Some features of preferred/dispreferred turn shapes. In: (J.M. Atkinson and J. Heritage, eds) Structures of Social Action. Cambridge University Press. Cambridge, pp. 57-101.

Reisman, K. 1974. Contrapuntal conversations in an Antiguan village. In: (R. Bauman and J. Sherzer, eds) Explorations in the Ethnography of Speaking. Cambridge University Press. Cambridge, pp. 110-124.

Rinott, M. 2008. The laughing swing: Interacting with non-verbal human voice. Proceedings of the 14th International Conference on Auditory Display, Paris.

Sacks, H., Schegloff, E. A., Jefferson, G. 1974. A simplest systematics for the organization of turn-taking for conversation. Language 50:696-735.

Schegloff, E. A., Jefferson, G., Sacks, H. 1977. The preference for self-correction in the organization of repair in conversation. Language, 53:361-382

Schegloff, E.A. 2000. Overlapping talk and the organization of turn-taking for conversation. Language in Society 29:1–63.

Scollo, M. 2011. Cultural approaches to discourse analysis: A theoretical and methodological conversation with special focus on Donal Carbaugh's Cultural Discourse Theory. Journal of Multicultural Discourses, 6: 1–32. doi:10.1080/17447143.2010.536550

Scollon, R., Scollon, S., 1981. Narrative, literacy, and face, in interethnic communication. Ablex. Norwood (NJ)

Sidnell, J. 2001. Conversational turn-taking in a Caribbean English creole. Journal of Pragmatics 33(8):1263-1290.

Stivers, T., Enfield, N. J., Brown, P., Englert, C., Hayashi, M., Heinemann, T., Levinson, S. 2009. Universals and cultural variation in turn-taking in conversation. Proceedings of the National Academy of Sciences, pp. 106-126.

Stokes, R. and Hewitt, J. 1976. Aligning Actions. American Sociological Review 41:838-849.

Tannen, D. 1993. The relativity of linguistic strategies: Rethinking power and solidarity in gender and dominance. In: Gender & Discourse. Oxford University Press. New York, pp. 19-52.

Tannen, D. 2012. Turn-taking and intercultural discourse and communication. In: (C. B. Paulston, S. F. Kiesling, and E. S. Rangel, eds) The Handbook of Intercultural Discourse and Communication. Blackwell Publishing, pp. 135-157.

Wang, P., Winter, U., Grost, T. 2015. Cross cultural comparison of users' barge-in with the In-vehicle speech system. In: (A. Marcus, ed) Design, User Experience, and Usability: Interactive Experience Design. DUXU 2015. Lecture Notes in Computer Science, 9188. Springer.

Wechsung, I. 2014. An evaluation framework for multimodal interaction: Determining quality aspects and modality choice. Springer International Publishing. Switzerland.

6 "I had already started blabbing" – User Barge-in

6.1 Introduction

One of the areas that proves to be of interest for the study of cultural differences in human-machine communication is interactional misalignments, or moments of confusion, misunderstanding, etc. This chapter focuses on one potential indicator of such misalignment, user barge-in behaviour, or in the context of speech systems an occurrence in which the user starts to speak before the speech system prompt has finished, thus creating a temporary overlap of talk.

It has long been established that the distribution of participation and speakership in conversation underlies orderly practices known as turn-taking organization (Sacks, Schegloff & Jefferson, 1974). In a nutshell, people manage turns by one party talking at a time. However, certain situations may allow for verbal gaps or overlaps in conversation. An empirically grounded account of these situations, their classification and set of practices to resolve overlap situations was provided by Schegloff (2000).

Most research on "barge-in" as a feature in speech applications revolves around technological challenges (Ström & Seneff, 2000; Rose & Kim, 2003; Raux, 2008; Selfridge et al., 2013). However, "barge-in" behaviour has long been linked to human turn-taking organization (Heins et al., 1997). Selfridge & Heeman (2010) provide an interesting overview on turn-taking conventions between users and speech systems which in turn informs system design. For turn-taking to be successful between a human and a speech system, the system should be designed to play prompts at appropriate times to create easily recognizable transition relevance places (TRPs) and subsequently start listening to the user response. In order to achieve this, the system has a predefined set of parameters. Such parameters define the system's turn-taking organization, including, for instance, user response time before it decides that the user hasn't responded, as well as start and end of speech detection. If users do not formulate their utterances within these time frames, then the dialog suffers from interactional misalignments. Consequently, a speech system can only support user "barge-in" if it begins listening for user speech during the playback of prompts.

Based on Wang, Winter & Grost (2015), in this chapter we discuss various user "barge-in" instances in the US and China to better understand in which situations users accidentally or intentionally "barge-in," and subsequently try to resolve a resulting overlap. We pay attention to the users' expectations of system capabilities and behaviour in these occurrences. And we analyse the different normative expectations shown by participants in both cultures, as well as the design implications for in-vehicle speech systems.

While we recognize the term "barge-in" is often employed among HMI researchers, from here we adopt a less morally infused conceptualization of the

https://doi.org/10.1515/9783110519006-006

phenomenon as "overlap," since the status of overlapping speech as a "barge-in," or not, and whether it is treated as such by human users, is an empirical matter born out of analysis of any particular instance.

6.2 User Overlap in the US Data

While the Wizard of Oz system is capable of listening for a user-initiated overlap, studies in the US and China did not use this feature. Therefore, in all occurrences of overlapping speech the system continued to play its prompt. The participants' behaviour in such cases must be analysed and viewed under this constraint. In the US study, the first 4 participants were part of a pilot study and as a result are excluded from the analysis here. Hence, the data from participants No. 5 through No. 26 are the basis for the analysis.

Only 6.54% of all 596 dialogs contain a user-initiated overlap in one turn of a communication event, altogether 39 occurrences. These instances are distributed unevenly over participants. Three participants overlap with the system for around 25% of their dialogs (between 5-9 times, P7,9,17). Another 4 participants have 2 or 3 instances of overlap (P6,10,14,24). All other participants do not initiate an overlap with system prompts at all, or do this once. In our analysis, when the user interrupts the system prompt, the square brackets [ ] indicate the start and the end of the overlapping speech. Our field data included the following communication event:

Instance 1
Participant 22 - 50:38.20
P = Participant, S = System
1 P: ((touches microphone button and system dings)) Call T-.
2 S: Which number for T- ? [Home, work?
3 P: [Home
4   (3.0)
5 P: Home

In line two, in between the question "Which number for T-?" and its subsequent listing of the possible phone types "Home, work?", the participant responds to the speech system by answering the first question with "Home". When he hears that the system prompt continues, he repeats his answer after the prompt has finished. Similarly, we can observe in the next two examples that the participant overlaps with the system prompt after the end of its first sentence, and before the second sentence, likely indicating the user's sense that the prompt had finished.

Instance 2
Participant 9 - 28:23.20
1   S: Sorry. I did not understand your request. [Please repeat it again
2   P: [Fox-
3   (3.0)
4   P: F- Fox news

Instance 3
Participant 7 - 41:50.5
1   S: Please review your specified artist. [It does not appear in your
2   P: [Ok
3   S: collection.
4   (3.0)
5   P: Ok, let's do something easier ↑Bruce Springsteen↑

Of the overlaps observed in the data, 16 instances follow the above pattern where a user appears to believe the system's turn is over and provides information that is responsive to the system's inquiry. However, the misalignment becomes clear when the system continues to elaborate the prompt, adding an unexpected phrase afterwards. In all of these occurrences the participant drops out, returns the floor to the system, and waits for the end of its prompt. Only when they are sure that the system indeed has selected them as speaker, the attempted user turn is reinitiated and completed.

The next two examples, however, show a slightly different situation, as the system has not yet indicated a transition relevance place, either through terminal intonation or pause, when the participant initiates an overlap.

Instance 4
Participant 17 - 19:41.60
1   S: Which station [or channel do you want to hear?
2   P: [X-
3   (2.5)
4   P: XM

Instance 5
Participant 17 - 21:19.70 (immediately repeats same request)
1   S: Which station [or channel do you want to hear?
2   P: [e-
3   (2.6)
4   P: XM

In two sequential instances participant 17 starts his utterance after the system begins its prompt "Which station." This in itself could be a sufficient system prompt as it contains all information necessary for the user to request a station name or number. Only the intonation, which is neither particularly falling nor rising, hints to continuation of the prompt. As observed for the other participants, participant 17 interrupts himself immediately and gives the floor back to the system till the end of the prompt. An additional example of the phenomenon can be found in the following instance from participant 24.

Instance 6
Participant 24 - 21:06.40
1   S: What music channel [do you want to hear?
2   P: [live jazz please

A total of 9 instances exhibit his pattern of overlapping. In it, the system does little to indicate a point of transition through either terminal intonation, a pause, or where grammatical completion is possible. The context of these instances is similar to the previous in that the system has already grounded the user on its expectations at the point of overlap so that the user can take his turn. Claiming with any confidence whether users understood the system's turn to be complete and thereby initiated a response, or whether they initiated an overlap in order to try and achieve the completion of their request without having to listen to the remainder of the prompt is unclear. A case could be made that the combination of car noise and the system's artificial intonation voicing may lead them mistakenly to believe the system has finished its turn. What is clear, is that participant 17 initiates their turn at a potential point of grammatical completion for the system, but also knows from previous instances that the system's prompt is not yet complete. That the participant cuts themselves off in both instances when the system fails to abandon its turn may suggest that the participant was simply misaligned with the system's projected turn completion more than once. Together the above patterned instances make up 64.1% of all observed user-initiated overlaps.

A number of instances exist where the user's overlap with the system either occurs at a place where the user does not yet have enough information to be able to offer the system a response that is relevant to the system's query, or where the users overlap consists of a command that is not responsive to the system. These instances suggest that at times users may intentionally overlap the system's speech out of a desire for expediency, rather than wait to initiate an action after the system is finished speaking. This pattern can be identified in the following instances.

Instance 7
Participant 14 - 1:09:06.20
1   S: What [kind of music do you want to hear?
2   P: [next track

Instance 8
Participant 17 - 1:21:19.9
1   S: What music station [do you want to hear?
2   P: [End music
3   (4.9)
4 P: Stop

In these instances, while the user response is very short it is nevertheless overlap that is sustained until turn completion rather than cut-off as in the prior instances. In both instances the participants don't repeat their request but wait to see if the system has listened while playing its prompt. Since the users clearly don't intend to respond to the content of the system's prompt it is reasonable to conclude that their overlap is intended to intervene in the system's ongoing turn and shift action prior to the system's turn completion.

While the content of the user overlap in the above instances includes a system command, or an attempt to respond to the system's request for additional information, some overlap with the system includes talk designed as a backchannel confirmation that the system neither needs nor has requested. This patterning can be found in 7 instances in the corpus and can be seen in the instances below.

Instance 9
Participant 13 - 37:23.80
1   S: What station?
2   P: 99.3
3   S: Tuning radio to 99.3 [FM Laser 99.3 WLZX
4   P: [Ok
5   (6.8)
6   P: There you go

Instance 10
Participant 7 - 41:07.60
1   S: OK. Let us cancel the dialog. [You may request help at each step of
2   our interaction.
3   P: [OK.
4   (5.5)
5   P: OK.

According to Schegloff (2000) backchannels are excluded from the "one-speaker-at-a-time" practice, because the listener's intention is not to assume speakership but rather to support the ongoing speaker in his turn and as a result can carry relational meanings.

Finally, there are 2 instances of overlap that are difficult to easily identify as patterned in the same way as the above. In the first instance included below the user initiates a voice command, waits long enough for the system to begin speaking and then attempts to answer what appears to be a projection of the system's completed turn informed by past experience with this kind of exchange. However, when the system fails to yield the floor, rather than electing to finish providing the system with the information they believe likely to be requested, they cut-off the overlapping speech and try again, only getting one speech sound further to providing the system with the information it will need to act on their request. After two false starts, the user eventually completes the utterance after the system has passed the turn.

Instance 11

Participant 17 - 22:32.70

```
1   S:  Which [station [or channel do you want to hear?
2   P:         [e-      [ex-
3       (1.7)
4   P:  XM radio
```

It is hard to suggest at this point, after having gone through this sequence with the system twice before that the user does not understand that the system has not reached the end of its turn. More likely the user knows precisely what the system will ask in this sequential position, and desires to skip ahead to their turn in the hopes that system will forgo the remainder of its own and act on the provided information immediately. If this is the case, however, why start an intentional overlap and then abandon it? This may be because the user is hoping that when the system hears his initiation of an overlap that it will abandon its turn and listen exclusively to the user. When it becomes apparent that the user's bid for the floor is unheard or rejected, the user initiates a second bid, which is also not picked up by the system.

The second non-conforming instance appears similar to instances where users issue a complete command during the system's turn, and not at a discernable TRP. This instance is distinguished however, by the length of the sustained overlap, as in other instances of this type the issued command is short, often comprised of two words, which this instance requires a significantly longer sustained overlap, or what Schegloff (2000) referred to as a competitive co-production. The irony of this competition is that the system is incapable of stopping, but also cannot project the oncoming end of a user's talk and hence is unable to "outlast" a determined user who can project the system's oncoming turn completion. This dynamic is discussed more thoroughly in Chapter 5.

Instance 12
Participant 7 - 47:20.90
1 S: Please let me [know what you want.
2 P: [Make a phone call
3 P: Make a phone call.

Despite the number of users who struggled with overlaps with the system, only participant 12 noted the trouble in their interview, saying "she would say 'What station do you want', but I had already started blabbing". In other interviews, there are occasionally mentions about the length of the prompts, which may relate indirectly back to perceived transition relevance places in the prompts. Participant 17 says, "I would also think that some of those commands could be quite a bit shorter. 'Which XM station or channel do you want to hear?', that's almost TMI." Others find the timing between the user and system turn taking unnatural and arduous, as participant 16 points out:

> yeah, I mean, that's where ( )... that's where the whole personable aspect sort of failed, because the idea is to sort of make it feel like it's natural and you're interacting with a person, it just sort of like... it's like one of these things when you're talking on the phone and there's a delay, and so you both like starting and stopping and you don't know when to talk and when the other person is listening.

In summary, our observations indicate that there is a difference in how users react when they overlap with the system but the prompt does not stop immediately or does not stop at all. In those 26 instances (66.6%) when participants become misaligned with the system and appear to believe that a TRP has occurred, they mostly cut-off and then repeat the utterance at the end of the system prompt, which is an expected human repair mechanism to resolve interactional misalignments (Schegloff, Jefferson & Sacks, 1977; Schegloff, 2000). Perhaps this cut-off rate may have even been higher had the length of time a user would need to hold the floor in order to complete their overlapping turn was greater. On the other hand, in 13 instances (38.4%) participants seem not to be irritated by the continuing prompt and don't cut-off or repeat their speech, appearing to believe that the system can and has heard them. Most of these instances involve users providing backchannels, or switching/ending tasks, which are not dependent on comprehension of the complete system prompt. In such cases the overlap is designed to complete the task at hand faster by attempting to bypass the remainder of the system's turn.

6.3 User-initiated Overlap in the China Study

In the China study, one participant showed inexperienced driving skills and the experimenter decided to terminate that driving session. Data of the remaining 25 participants are used for the analysis. There are 97 instances of overlapping speech at 11.8% of the 822 communication events in the China study. The instances are distributed more evenly over participants than those in the US data. Approximately half of the participants initiate an overlap for around 20-25% of their dialogs (between 5-9 times). Most of the remaining participants have 2-4 instances of overlap. Only 4 participants do not overlap system prompts at all or do this only once (P2,7,14,24).

   As in the US data, there are many scenarios in which participants overlap likely because of a misalignment with the system on what constitutes a transition relevance place, or because the user has enough information to proceed with their next action and intentionally overlaps the system in an attempt to move the interactional sequence forward more quickly. Here is one example, in which the participant self-interrupts and repeats his utterance after the prompt has finished.

Instance 13
Participant 13 - 41:27.30
1   S: Which album would you like to listen to? *您想听什么专辑?*
2   P: Black 。
3   S: Please confirm Black [and White. *确认 Black and White。*
4   P: [Con- *确。*
5   P: Confirm. *确认。*

In the next example the system prompt has provided the user with the information necessary to proceed in a way that was also observed in the US data:

Instance 14
Participant 8 – 19:22.60
1   S: Which radio station [would you like to listen to? *您想听[什么]电台?*
2   P: [End- *结*
3   P: End broadcast. 结束广播

Participants frequently overlap when the system prompt includes a list of items, such as available songs or artists. During the system prompt, users may not wait until the end of the prompt but instead overlap after hearing their desired item from the system, as in the following.

Instance 15

Participant 18 - 1:03:05.20

1   S: What music would you like to listen to? *您想听什么歌？*
2   P: Tell me the songs by Legend of Phoenix. *报一下凤凰传奇的歌*
3   S: ((system begins verbal list of song names))
4   ...
5   Lo[ve someone who doesn't go back home, I believe...
6   P: [Love someone who doesn't go back home. *播放爱上一个不回家的*
                 人

While this is a frequent scenario in the China data, only participant 9 in the US data overlaps with a backchannel signalling confusion, but otherwise listens to the complete list of radio stations before taking his turn.

Another type of overlap that doesn't occur in the US data includes participants taking their turn during system prompts, which are sometimes reported as being too long with an overwhelming amount of choices. This is one area where users make use of the multimodal capacity of the system to choose, for instance, from a possible list of albums displayed on the center stack. With a short glance, rather than listening to a lengthy verbal prompt, a user can choose the album from the list.

Instance 16

Participant 13 - 31:31.50

1   S: Which artist would you like to listen to? *您想听谁的歌？*
2   P: Liu Dehua *刘德华。*
3   S: Play Liu Dehua or select an [album. *播放刘德华或选择一张[专辑。*
4   P: [The classic restored. *经典重现。*

Additionally, the time provided to make a selection after the end of the prompt is relatively short, after which the system will take the default action of playing a random selection from the artist. Consequently, if a participant wants to select an album they may learn that this selection needs to be made quickly to avoid the default action being accepted, and begin to select an album before the prompt is complete.

The next instance shows a participant attempting to switch task. When this is the action being sought the system's prompt becomes irrelevant, only delaying the user's ability to accomplish the action they desire. Given this, we might speculate that a user who waits for the system's turn to complete does so either because they don't believe the system is able to hear or act on the content of their overlapping speech (the system isn't listening), or perhaps the norms that often guide human interaction that yields some rights to the speaker to complete their turn are also being extended to machines. In the following instance a user engages in overlap in the process of a task switching event.

Instance 17

Participant 22 – 1:06:44.8

1 S: Who would you like to call? 您想给谁打电话？

2 P: Mr. Dai. 戴先生。

3 S: Please [wait. 请稍等一下。

4 P: [Hey, I don't want it anymore, I don't want it anymore, I don't want it anymore, I don't want to make a phone call anymore 哎，不要了，不要了，] 不要了，不想打电话了。

5 P: I want to navigate, can I navigate? 想导航了，能导航吗？

In the following instance, the participant takes time to make his selection from a long artist list. While the participant scrolls the list manually, the system waits for his response. But when it still takes some seconds to make the choice after the last manual scroll operation, the speech system decides that the user's turn has timed out and self-selects as next speaker to remind the participant to choose an option. The participant overlaps the system's reminder with their choice almost simultaneously and repeats their choice after the end of the prompt.

Instance 18

Participant 13 - 53:02:9

1 S: Which artist would you like to listen to? *您想听谁的歌。*

2 P: ((quickly scrolls down artist list by touch)) (10.0)

3 S: Please [let me know what you would like to do. *请告知您想做什么。*

4 P: [Legend of Phoenix. *凤凰传奇。*

5 P: Legend of Phoenix. *凤凰传奇。*

From the instances above we can see that often the prompting circumstances for overlap are similar to the US data. Participants may try to advance the dialog faster, and thus overlap when the information provided by the system is known or irrelevant to their next action, or if they are misaligned with the system on the constitution of a transition relevance place.

From all 97 barge-in instances, in only 12 instances (12.4%) users hesitate or cut-off after an overlap during the ongoing prompt. In 38 instances (39.2%), users complete their utterance during the ongoing prompt and repeat it after the system completes its turn. In the remaining 47 instances (48.5%) participants overlap during the system prompt and do not repeat, likely presuming they have or will be heard despite the overlap. That these overlaps during system prompts occur more frequently in the Chinese data overall, as well as more specifically when users seek to move the ongoing turn forward by overlapping the system with a new command, may suggest that some Chinese users do not operate in accordance with a norm in the US data that yields some rights to the system to complete its turn without interruption.

6.4 Cultural Variability in Overlap Timing

Because the speech processor in this study is not the system, but a human "Wizard" serving this function unbeknownst to the participants, the speech overlaps identified above are handled with more flexibility than a real speech system would currently be capable of, in part by applying human overlap resolution practices. In a real scenario, most of these cases would lead to misrecognition or recognition with lower system confidence, so that in fact this user-initiated overlap behaviour likely causes subsequent system initiated overlap when the system starts a new prompt while the speaker repeats his utterance. This may also result in confirmation or repetition requests, thus extending the number of turns and overall dialog time. In this context, an important observation is the time frame from the beginning of the user utterance until the user interrupts himself. In the US data, user 17 interrupts himself very fast between 130 and 180ms, while the prompt continues. All other participants stop speaking only after 400-500 milliseconds. The data collection does not have enough instances to conclude a minimal time frame, but it seems that most users would speak for at least 400ms before interrupting themselves. Further investigation is necessary.

Another important observation is the time frame in relation to a mistakenly perceived transition relevance place, which leads to design recommendations for an overlap accommodation feature. Frequently, misjudged transition relevance places occur when the system makes a pause between phrases or sentences, or reaches a point of possible grammatical completion. The average time frame between the end of the last system word and the beginning of the user response is 400-500ms. Only one participant in the US data and one in the China data starts to talk at 290 and 310ms respectively. These observations are used to inform design implications in the next sections.

6.5 Cultural Variability in Norms for Interaction

Socio-culturally speaking, the US data suggest that overlap behaviour violates cultural norms, except under certain conditions. This can be clearly derived from the fact that most overlap instances occur when the user is misaligned with the system on the appearance of a transition relevance place and self-selects as next speaker. Most frequently, US participants abandon their turn when the overlap becomes apparent and the system does not move to abandon it's turn. The participants then wait until the next transition relevance place when the system yields the interactional floor to reissue their command or provide information the system appears to be requesting. This pattern is diverged from only when a user initiates a task switch as the remainder of the system prompt is then irrelevant to their next action. When this occurs some users initiate their command in sustained overlap with the system, and occasionally don't move to repeat the command, apparently assuming the system can and did hear the initial command even through the overlapping speech.

The apparent preference in US data to generally avoid overlap unless interjecting to change the course of the interaction can be explicitly formulated in the following interactional norm:

1. *When interacting with another, it is preferred that utterances not overlap, unless interjecting to change interactional trajectory.*

And further:

2. *If utterances overlap, one should yield the floor to the speaker until their turn is complete, unless interjecting to change interactional trajectory.*

Additionally, we might formulate a cultural premise that appears to inform this normative behaviour in the cultural context of the car that: *1) Respect for the speaker's right to hold the floor until yielded also extends to non-human interactants.*

Our data from China suggest that norms governing overlap practices do not require users to avoid overlap as persistently. While in both data sets participants use overlap as a way to speed up the interaction, particularly when the remainder of the prompt is projected to be irrelevant, there are many more overlap instances in our data from China. These overlaps include a variety of situations identified throughout the chapter, which do not lead to user-initiated overlap in the US data. Users often overlap during the prompt at places which have not reached a potential point of grammatical completion, nor employ a pause indicative of turn exchange, even when considering potential differences in what length pause may constitute a TRP cross-culturally. This is further supported by our observations that Chinese users more frequently talk over the system prompt once they are able to project the completion of the turn and have enough information to provide a response. This is something that US users appear to prefer not to do. The result of this willingness to initiate an overlap as soon as they are able generates more instances of sustained overlap in Chinese data, while US users mostly abandon their turn and wait for the system to finish the prompt. This pattern persists among Chinese users even after multiple instances. This suggests that the cultural variation observed here cannot be accounted for on the basis of confusion about whether the system is able to hear and act on overlapping speech.

Our findings are supported by another difference in both data sets. We have measured the user response time for all participants' utterances in both data sets excluding the overlap instances. User response time is defined as the elapsed time between the end of the system prompt till the beginning of the user utterance. The response time of Chinese users is in the average 1.27 seconds (arithmetic mean) with a standard deviation of 0.85 seconds. In the US data the user response time is 1.58

seconds in the average with a standard deviation of 1.45 seconds. This is statistically significant (Sig<0.001). The significance also holds true when looking only into those utterances where a participant is absolutely certain about the system's capabilities, and his own intention at the time of his turn taking. In this case, the average user response time in the US data is 1.24 seconds with standard deviation of 1.1 seconds versus an average of 0.92 seconds with deviation of 0.61 seconds in the China data. Similar to the higher number of instances of overlap, Chinese participants respond faster to the speech system, and in interviews indicated their assessment that the system is often "too slow." While this generally faster response time does not directly inform overlap practices, it may indicate a preference among Chinese users for speed and efficiency of interaction over allowing the system to complete it's turn at the expense of overall time to task completion, or time spent in dialogue.

These findings are consistent with an orientation to the system in China that is elaborated in chapter 4. In this orientation Chinese users act toward the system on the basis of verbal depictions of the relationship between user and system as one of an employer to employee, or more generally, a service orientation. In the US data, the system was either oriented to as a kind of simple machine or as a potential friendly interactional partner, on near equal interactional footing.

Given these differences in orientation to the system the observed findings of Chinese users being more willing to overlap with the system in order to have their needs met as quickly as possible is complimentary to an orientation to the system as primarily there to serve the needs of one who is hierarchically superior. The observation that US users are more likely to yield the floor to the system and abandon overlaps when the system does not drop out immediately suggests a more egalitarian orientation to the system that respects the system's rights to complete it's turn at talk. That most US instances of overlap occur at moments when misalignment between the user's and system's understanding of a potential TRP suggests that generally users are not trying to overlap the system in order to speed up interaction, and that generally instances of overlap are accidental and quickly repaired. That many instances of overlap in the Chinese data do not occur at these points of potential misalignment and are more frequently sustained in a competitive co-production suggests Chinese users' sense that they are entitled to overlap the system in service of speed and efficiency and that the system ought to abandon its turn in favour of hearing and acting on the user's command as quickly as possible.

As a result of this analysis we can formulate the following Chinese interactional norms:

1. *When interacting with another who is hierarchically inferior, it is permissible that utterances overlap.*

And further:

2. *If utterances overlap, the hierarchically subordinate interactant should yield the floor.*

An accompanying cultural premise can also be formulated that: 1) Hierarchical deference expected among human interactants also extends to non-human interactants.

6.6 Design Implications and Conclusion

6.6.1 Importance of Overlap to Users

In the US and in China overlap, if intentionally initiated by users, seems mainly to serve the purpose of advancing task completion more quickly, though the US data strongly indicate that in most contexts users prefer to wait for a perceived transition relevance place to initiate a turn exchange. A system that was designed to respond to user's overlapping speech by cutting off its own turn may not be a feature that is immediately utilized because of this normative preference. As it operated in this study, it is possible that the system's seeming refusal to cut-off it's turn may have furthered user expectations that overlapping the system is dispreferred. In order to normalize the use of overlapping speech as an acceptable bid for the floor so that users might make use of a "barge-in" feature the system needs to provide some feedback signalling that the user is allowed and encouraged to do so. This might take the form of ending the system prompt immediately when user speech is detected. If despite user overlap the system continues the prompt, users may understand that they need to repair a norm violation, e.g. by repeating and avoiding further overlap behaviour.

The China data reveal that Chinese users do not necessarily perceive overlap as a violation of a communication norm, at least in the contextual setting of human-machine dialog in a vehicle. Chinese users in this study do show a significantly higher tendency to overlap system prompts, at places without transition relevance, as well as to continue to talk in parallel to the speech system, seemingly expecting the system to abandon their turn, and believing it appropriate that the system should do so. As a result, for these Chinese users, full support of a "barge-in" feature may be of higher value than for US users.

6.6.2 Prompt Timing and Wording

The data provide insights for recommendations on the prompt design itself. Designing short phrases or sentences has the advantage of creating a scenario less conducive to misalignments that may produce unintentional overlap. On the other hand, such short prompts have the disadvantage of not offering enough guidance for the user to provide the right amount and kind of information to the system in response, which ultimately hurts the overall interaction time that some users who overlap may be seeking to decrease.

Prompts that exceed the length of one short sentence or phrase should avoid potential grammatical completion, other than at the end of the prompt, as well as pauses that might indicate a transition relevance place. If a prompt cannot avoid this, pauses between parts of sentences and at any potential transition relevance place need to be significantly shorter than 400ms to prevent turn-taking misalignments. As seen above, when users take their turn but the system does not stop fast enough, users likely enter into a repair sequence which poses difficulties on the speech recognizer. Further studies indicate that pauses shorter than 120ms are effective. Such prompts should also avoid a terminal or rising intonation that suggests the end of a question or statement. The majority of users indicate in interviews that they would appreciate shortened prompts that provide the minimal feedback necessary for task achievement (e.g. "which radio station?", "What's the street?").

The recommendation to shorten prompts is supported by our numerous observed cases where users appeared to feel they had received enough information to respond long before the prompt finished (e.g. "What station [or channel do you want to hear?"). For example, in the Chinese prompt "Call which contact number? 呼叫哪个 联系人号码？" the element "number" is redundant and user misalignments on the status of a transition relevance place would be less likely without said element. In another case, a user requested "Black", which is part of a song name, so when the system asked for confirmation, the user started to confirm immediately after the system said "Black." The confirmation prompt could be designed as "Black and White, right?" allowing for overlap after the word Black to enable the user to provide the confirmation at any point.

Yet another type of prompt with multiple perceived transition relevance places is read-aloud prompts, where the system reads out lists of artists or radio stations upon user request. Rather than wait for the remainder of a list of irrelevant options to be read aloud, users would generally like to be able to overlap the system and confirm immediately when they hear their desired item.

6.6.3 Proposed Solution

Even though the data collected from the US and China are culturally distinct to the extent described in this study, both data sets support that turn-taking, and hence overlap, is negotiable between dialog partners. Since the seminal work by Sacks, Schegloff & Jefferson (1974), the negotiable nature of turn-taking has been studied and is well-established. In this light, a dialog system may not always want to allow the user to take the floor and self-select as the next speaker. This is the case for instance if the remaining prompt still contains information necessary to progress towards task completion, e.g. "I can't find a phone. Please connect your phone and try again."

Additionally, some users tend to express agreement through backchannels, such as "OK". In these or similar situations, a full "barge-in" feature may stop a prompt as the system cannot distinguish between continuers and bids for the floor. We propose that the system design should allow "barge-in" only in certain places that are likely candidates for user-initiated overlap, or alternatively once the system has conveyed the most important information. For example, in a prompt such as "I'm sorry, I didn't understand. What did you say?" Overlap most likely occurs from the last syllable of the first sentence through the first word of the second sentence: "I'm sorry, I didn't understand. What did you say?" On the other hand, a prompt such as "Which contact name would you like?" can allow for "barge-in" right after the word "contact", because users know what they need to say to complete the task. A dialog system that better understands how and when users project turn completion could be designed to allow "barge-in" in those places, while not responding to it at other times. Such an approach gives the system the ability to negotiate turn taking in a way that more closely resemblances human interactants attempts to bid for the floor that are sometimes taken up, and sometimes rejected. While potential misalignments are still possible, particularly if users are unable to discern why/how the system sometimes allows overlap and sometimes doesn't (important grounding work), we believe the benefit of providing users as much control over dialogue flow as possible provides maximal flexibility for applications across cultures as such a system would enable users who desire to move the interaction forward more quickly will be able, but not be required to do so.

References

Heins, R., Franzke, M., Durian, M., Bayya, A. 1997. Turn-taking as a design principle for barge-in in spoken language systems. International Journal of Speech Technologies 2(2):155-164.

Raux, A. 2008. Flexible turn-taking for spoken dialog systems. PhD Thesis, CMU.

Rose, R. C. and Kim, H. K. 2003. A hybrid barge-in procedure for more reliable turn-taking in human-machine dialog systems. In: Automatic Speech Recognition and Understanding. ASRU'03. 2003 IEEE Workshop. pp. 198-203.

Sacks, H., Schegloff, E. A., Jefferson, G. 1974. A simplest systematics for the organization of turn-taking for conversation. Language 50:696-735.

Schegloff, E. A., Jefferson, G., Sacks, H. 1977. The preference for self-correction in the organization of repair in conversation. Language, 53:361-382

Schegloff, E.A. 2000. Overlapping talk and the organization of turn-taking for conversation. Language in Society 29:1–63.

Selfridge, E. O., Heeman, P. A. 2010. Importance-driven turn-bidding for spoken dialogue systems. In: Proceedings of the 48th Annual Meeting of the Association for Computational Linguistics. Uppsala, Sweden, pp. 177-185,

Selfridge, E. O., Arizmendi, I., Heeman, P. A., Williams, J. D. 2013. Continuously predicting and processing barge-in during a live spoken dialogue task. In: Proceedings of SIGDIAL, Metz, France, pp. 384-393,

Ström, N., Seneff, S. 2000. Intelligent Barge-in in Conversational Systems. In: Proceedings of ICSLP. pp. 652-655.

Wang, P., Winter, U., Grost, T. 2015. Cross cultural comparison of users' barge-in with the In-vehicle speech system. In: (A. Marcus, ed) Design, User Experience, and Usability: Interactive Experience Design. DUXU 2015. Lecture Notes in Computer Science, 9188. Springer.

7 Apologies in In-Car Speech Technologies

7.1 Introduction

Human-like behaviour and social cues in in-car speech systems constitute a rich de-sign area, as these elements evoke perceptions of personality traits that may overtime lead to increased likeability and trust. Such human-like speech responses accomplish various self-presentational goals and touch upon different expectations in terms of the relationship users build with their in-car systems. In general terms, within an in-car communication event, practices that orient toward relational aspects rather than purely task-oriented turns become salient, as they may come across as unexpected and outside the canonical strategies so far used by speech enabled technologies, thus adding a surprising and delightful layer to the user experience. In this chapter, we delve into strategies to convey apologies as key relational elements that may increase phatic communication during the user-vehicle interaction. Drawing on user-car inter-actions and interviews from two user studies, we analyse participants' orientations toward the system's apologies. In particular, we discuss expectations of compliance to communication norms when the system is perceived either as a mere tool, or rather as a social tool.

7.2 Theoretical Framework and Related Literature

7.2.1 Culture and Communication Practices

As noted in the introductory chapters of this work, we view the car and the interac-tions with the in-car speech system as a specific "communication situation", follow-ing the tradition of the Ethnography of Communication (e.g. Hymes, 1972) and Cul-tural Discourse Analysis (Carbaugh, 2007; Scollo, 2011). From this key concept, we derive that the car is a discursive space where people communicate with each other and with the car in cultural ways. Culture, in this use refers to "a historically trans-mitted expressive system of communication practices, of acts, events, and styles, which are composed of specific symbols, symbolic forms, norms, and their meanings (Carbaugh, 2007, p.169). As a result, culture is understood as both constituted by, and constitutive of, discursive practice. Therefore, it is illuminated by analyses of social interaction as interlocutors express their understanding of the premises and norms that inform that interaction both in the culturally distinctive ways it is performed, but also in the taken for granted shared knowledge that renders this expression mutually intelligible.

Within the in-car communication situation, there are smaller sequences of acts that can be understood as "communication events" with each such event composed by "communication acts" (Hymes, 1972). Typical communication acts in our field data

https://doi.org/10.1515/9783110519006-007

are openings, directives, confirmations, and closings. Together with communication act sequences, the applied framework explores important extra-linguistic, multi-modal cues active in these communication practices, as well as the cultural premises that inform the production and interpretation of the meaningfulness of the practice, and the norms that regulate the moral dimension of their use. Cultural premises, in their conceptualization under CuDA, are analytical formulations that direct our attention to various semantic "hubs," rich with cultural meanings about personhood, relations, emotion, place, and communication itself, that are both presumed and enacted in discourse. Relatedly, as seen from the lens of interpersonal ideology, formulations of actions toward others and interpretations of other's actions are guided by expectations of personhood, relations, and communication that take place during social interaction (Poutiainen, 2015)

Another key concept we rely on is "conversational inference" (see Gumperz, 1982, 1992; Carbaugh, 2005), which uncovers how participants assign meaning to nonverbal cues within the on-going sequence of communication events using shared cultural means. In a multimodal conversational interface, multimodal means, such as visual cues on displays, play a significant role and so their design – and the ways participants use and interpret them – partly determines the success of a communication event (Levinson, 1997, Pearl, 2016).

7.2.2 Human-Interface Relationship

Apart from the key communication concepts described above, we draw on the notion of social presence to understand the social dimensions that come to play when humans interact with interfaces through speech. Social presence can be defined as the sense of "being with another" including primitive responses to social cues, simulations of "other minds" and automatically generated models of the intentionality of others. Similar to the notion of "being with others" (Goffman, 1959; 1963) or "phatic communion" (Malinowski, 1923), social presence is activated as soon as a user believes that an entity in the environment displays some minimal intelligence in its responses (Nowak & Biocca, 2003). In interactions with speech systems, social presence can be said to take place at intermittent moments in which users overlook the artificial nature of the experience and can immerse themselves in a natural conversation, relying on their instinctive use of communication patterns (see Rosenbaun et al., 2016 for a review).

Because of its implications for interaction, social presence has been identified as a key variable in building the relationship between a given technology and its users (Hassanein & Head, 2007; Kumar & Benbasat, 2006), affecting perceptions of user trust (Gefen & Straub, 2004; Hassanein & Head, 2007) and serving as an enabler for trust-building cues (Gefen & Straub, 2003). Research has shown that perceptions of social presence influence perceptions of system usefulness (Karahanna & Straub,

1999), enjoyment (Hassanein & Head, 2004), involvement (Fortin & Dholakia, 2005), and trust (Gefen & Straub, 2003; Gefen & Straub, 2004). We understand social presence to include the feeling of warmth and sociability conveyed through interaction with a human-like voice assistant, which may be affected by and accomplished through levels of relational talk produced by the system. In this chapter, we focus on apologies as a key type of relational talk.

7.3 Apologies in HMI

Extensive research has been done on apologies in human-human interaction and on how they are accomplished. Broadly speaking, these speech acts are realized through routinized patterns used to 'save face' with various degrees of effectiveness (Scher & Darley, 1997). Apologies serve as remedial work, designed to smooth over or recover from any social disruption caused by a normative violation (Blum-Kulka, 1997). For an apology to be legitimate, transgressors must recognize their wrongdoings, and explicitly show regret (Fraser, 1981). Five strategies for apologizing have been suggested : (1) an illocutionary force indicating device (IFID; such as, "I'm sorry"; (2) an explanation of the violation; (3) a display of responsibility for the fault; (4) an offer of compensation; and (5) a promise of avoidance of wrongdoing in the future (Olshtain & Cohen, 1983; Blum-Kulka & Olshtain, 1984). Apology studies have shown that an apology draws favorable responses only when the genuineness of the apology is convincing (Skarlicki, Folger, & Gee, 2004). Furthermore, it has been argued that the use of just one apology strategy is more effective than the use of multiple strategies (Scher and Huff, 1991).

Within the HMI community, research has shown that when an interface has high autonomy, users are more likely to treat it as human and thus, the interface needs to live up to users' expectations in order not to be negatively assessed (e.g. Serenko, 2007). For example, an interface's modesty, in blaming itself and calling it an error in "the system," makes users feel better and more comfortable with the system (Nass, & Brave, 2005). However, when the interface engages in self-criticism, users perceive the interface as generally less competent (Park, 2015). These findings point to the delicate balance required when designing a modest yet competent voice assistant. Similarly, certain preferences of some apology strategies over others have been reported in computerized environments, indicating a strong correlation with preferences in social contexts, further supporting the notion that social dynamics that guide human-human interaction also apply to HMI (Akgun, Cagiltay & Zeyrek, 2010).

7.4 Data and Methodology

The field data analysed in this chapter have been referenced and discussed in previous publications (e.g. Carbaugh et. al., 2016, Molina-Markham et.al., 2016) and consist of interactions and interviews from two studies conducted in 2012 in the area of Amherst, Massachusetts, and in 2014 in the area of Columbus, Ohio. We label these two datasets "Study 1" and "Study 2".

Our analytical framework is informed by attention to the 8 basic components of the Setting, Participants, Ends, Acts, Key, Instrumentalities, Norms, Genre (SPEAKING) mnemonic proposed by Hymes (1972) that helps to describe, organize and interpret human behavior in cultural scenes (see Hymes, 1972; Carbaugh, 2012). We use the set as a conceptual foundation for an initial descriptive analysis, in which the components function as a series of questions that help to unveil how communication is done within and about the car. As a next step, these 8 basic components are used for interpretive inquiry.

In an additional step, design questions highlight HMI-relevant insights from the data. The design of an in-vehicle multimodal interface traditionally follows the paradigm of User-Centered Design (Norman and Draper, 1986; Nielsen, 1993; Vredenburg and Butler, 1996). As such, design dimensions identified and further explained in Carbaugh et al. (2016) are user satisfaction, user trust, ease of use, turn taking, grounding, cooperation, interaction style, multimodal use, conflict resolution, and information distribution across a communication act sequence, level of formality, among others.

Finally, interview data were analyzed by comparing the major themes that emerged as participants verbalized their experiences with the system and their various expectations toward voice assistants. This analysis framework was applied iteratively, as a series of frequent observations led to patterns and expectations that emerged from the data. In short, in the tradition of CuDA, we analyzed not only interaction data from communication occurring with the in-car system but also communication about the system, for evidence of cultural premises about personhood, relations, emotion, place, and communication itself, where relevant. We included our observations from HMI analysis as a complementary perspective to enrich and confirm the conclusions.

Data from both studies are included here to provide a sense of the range of orientations to communication in and with the car, and the potential cultural variation in the means and meanings employed in these interactions. We do not claim here to be presenting discrete cultural practices constitutive of a particular cultural community, but rather seek to explore the range of these practices, recognizing that all cultural discourses are "multidimensional, polysemic, deeply situated, and complex functional accomplishments" (Carbaugh, 2007, p.169).

7.5 Analysis

We would like to first frame our analysis by illustrating how some norms, premises, practices and strategies from human-human interactions are transferred to in-car interactions with speech technologies, a phenomenon we call "cross-polination." The following example shows how certain norms of politeness and strategies for correcting norm violations are displayed by participants' apologies to the system. Such apologies are non-task related turns that indicate users' adherence to a broader sense of cultural norms, and premises informing expectations of personhood, rights, and obligations.

Instance 1
Participant 13 – 34:15
P = Participant, S = System
1   P: (participant touches speech button) Call [H-
2   S: [(audible ding)
3   P: Oh, I'm sorry.
4   (6.1)
5   S: Please confirm calling H- S-.
6   P: Yes, call H- S-
7   (2.3)
8   S: Calling H- S- on mobile.

The participant initiates a new speech session by pressing the speech button and immediately phrasing his directive as a single turn ("Call H-", line 1). However, in the sequential organization of the interaction, pressing the speech button constitutes a turn-at-talk in itself and so, the speech system responds with an audible ding while the participant is saying the contact name "H-". Although it is the system that overlaps with the participant's intended turn, the participant orients to this overlap as being her responsibility, since the participant has already gone through various cycles in which the speech system responds with a ding when activated by the touch button, explicitly yielding the floor. The acknowledgement of responsibility by expressing mild surprise ("oh") is then followed by the apology "I'm sorry", immediately yielding the floor back to the system for further instructions on how to proceed.

As this example illustrates, and as discussed in chapter 5 of this work, human-machine communication is partly regulated by expectations derived from human-human communication, and deviations from such expectations are interpreted as having social meaning. In what follows, we explore the design tensions between apology strategies to convey a polite yet self-confident voice assistant.

7.6   Study 1

In the design for Study 1, the speech system apologized in two different types of situations. First, if users asked for a task the system could not accomplish, for example a request for route guidance that was not supported. Second, if the system repeatedly did not understand the user request. In both cases, the apology was short, either "sorry" or "I'm sorry", immediately followed by a brief explanation or account of the problem, an optional expression of the system's responsibility for the fault, and if possible, a follow-up request offered as a means to continue with the task. The instance below shows the first type of apology, produced when the participant asked for a music item not contained in the system music collection.

Instance 2
Participant 14 – 31:46
1 P: (participant touches speech button)
2 S: (audible ding)
3   (3.1)
4 P: Play John Lee Hooker.
5 S: Pardon?
6 P: Play John Lee Hooker.
7   (11.9)
8 S: Processing.
9   (11.3)
10 S: I am sorry. I am unable to find the music item that you are looking
11   for.
12 P: (shrugs shoulders)

The participant presses the speech button, which starts a new speech session while the system is in the home screen with no specific task at hand. The system responds with an audible ding, after which there is a 3.1 second pause (line 3), interpreted by the participant as his signal to take the conversational floor and utter a directive ("Play John Lee Hooker", line 4). The system immediately indicates a communication problem by asking the user to utter the directive again ("Pardon?" in line 5). Although this system response does not state the kind of problem at hand, it clearly signals that the request was not understood. The participant interprets the system's response as a misrecognition and repeats his directive verbatim (line 6). While the participant waits for the system response (lines 7 – 9), he glances multiple times at the display to understand whether the system is still engaged in the interaction. Since prolonged pauses in conversation are usually interpreted as indicators of an upcoming dispreferred response (e.g. McLaughlin & Cody, 1982; Pomerantz, 1984), the user may, borrowing from human-human interactional norms, interpret the system as having difficulties answering appropriately. The subsequent apology in line 10 ("I'm sorry") with the explanation "I am unable to find the music item that you are looking for" unsurprisingly confirms this communication breakdown, to which the participant simply shrugs his shoulders. About 10 seconds later, the participant initiates a new speech session and asks for a different artist (section not transcribed).

The second type of apology trigger is illustrated in the example below. Here, the system apology is a result of multiple subsequent instances of mishearing or misrecognition.

Instance 3
Participant 9 – 27:58
```
1   P:  (participant touches microphone button)
2   S:  (audible ding)
3       (2.2)
4   S:  What talk channel do you want to hear?
5   P:  Let's listen to something conservative.
6       (4.5)
7   S:  Pardon?
8   P:  Something conservative, right wing.
9       (4.3)
10  S:  Could you repeat that please?
11  P:  You know, fire and brimstone, that kind of stuff.
12      (4.8)
13  S:  Sorry. I did not understand your request. [Please] repeat it again
14  P:                                            [Fox-]
15  P:  F- Fox news
17      (7.2)
18  S:  Please confirm 98 XM The Fox xhole
19      (3.2)
20  P:  What's that? Fox hole? Ok
21      (5.8)
22  S:  Tuning radio to 98 XM The Fox xhole
```

At the time the participant initiates a speech session by touching the microphone button, the speech system's context is the XM Talk Radio context, and as per design the system triggers an XM Talk Radio related question ("What talk channel do you want to hear?", line 4). The participant does not request a specific channel number or name, but rather his directive is descriptive of the type of talk station he wants to hear (line 5, "Let's listen to something conservative"). What follows is a series of turns in which the speech system asks to repeat the directive after a 4-5 second pause (lines 7, 10, and 13), and the participant rephrases his directive, but continues to describe the preferred type of channel (lines 5, 8, and 11). In line 13, the third request for the command repetition, the system starts the response with a short apology to account for the multiple misunderstandings and takes responsibility for them. It explicitly states the reason behind the communication breakdown ("I did not understand your request") and follows up with an offer to continue ("Please repeat it again"). The participant changes his directive and asks for a specific XM channel "Fox news" (line 14). While the speech system still does not accurately recognize the command, it offers a similar sounding alternative, which the user accepts (line 18).

As these examples illustrate, the voice assistant was designed to take responsibility for the problem, whether the issue is triggered by the system's limited scope of tasks, or by the system's misunderstanding of a user utterance. As such, the apology serves the main purposes of grounding and hinting toward a task completion issue, letting the users know that their way of interacting with the system is not necessarily wrong. In this sense, the apology removes any potential user's uncertainty regarding how to interact with the speech system. Such short and concise apologies did not contain any markers of regret, and consequently, no emotion was evoked or alluded to by participants during the subsequent interviews.

7.7 "Does it apologize?"

As evidenced in the interviews, participants indicated feeling well-informed of how the interaction unfolded, and appeared to orient toward the apology in an unmarked manner, just as intended. In the various instances of apologies present in the dataset, the most common reaction was to simply continue with the task as seen in instance 2, or to initiate a new task as at the end of instance 1. Occasionally, participants would add a verbal backchannel, indicating their understanding or agreement, such as "ok" or "I know", or a gesture such as head nodding. In short, participants oriented to these apologies as discourse markers acting as indicators of issues within a task completion cycle. Because of this, the example below is the only mention across all participants regarding the system apologies.

Instance 4
Participant 7 - 1:42:10 – Interview
1 P: : I mean does it say I'm sorry? I didn't understand, does it say I'm-
2   (.) does it apologize?
3 I: Sometimes it does yeah. Did you hear that? ((Both laugh)) did you
4   like that it [apologized?
5 P: Well (.) I'm sorry (.) well, yeah (.) I don't know

This participant vaguely recalled the apologies when the system's level of politeness was discussed during the interview. This general absence of focus on system apologies seems to indicate that the apologies were perceived as natural and unmarked within the sequential organization of the interaction. The absence of any marker of regret, which can be part of canonical apologies, was not mentioned in the data, neither negatively nor positively. As observed in instances 1 and 2, all participants continued with the interaction after the system apology, strongly pointing toward the use of apologies for grounding purposes. Still unclear is whether apologies are heard and understood as sincere sayings, where the connotative meaning of apology is active vs. the more instrumental use of apology to mark problematic interactional moments for structural correction. The latter indicates a potential premise of personhood that does not grant the system the status of a full interactant and, as a result, without the same social responsibility of their human interactional partner. In study 2 below, on the other hand, more elaborate apologies conveying regret were employed.

7.8   Study 2

For the design in this study, the speech system was prepared to apologize essentially in the same two situations as discussed above, but in a more elaborated manner to add regret as illocutionary force, conveying hints of emotion. The following two instances show this system behaviour.

Instance 5
Participant 20 – 17:44:38
1   P: Wake up
2   S: (audible beep)
3   P: Call Charles Wilson at home
4   S: I don't have a home number for Charles Wilson. Would you like
5   mobile or work?
6   (3.0)
7   P: Neither
8   S: Pardon?
9   P: No
10 S: Sorry, I didn't understand. Please try again
11   (2.5)
12 P: End call
13 S: I'm so sorry, I still didn't understand. Clearly I'm limited. Please try
14   one more time.
15 P: Call mobile
16 S: Calling mobile

The participant enables the speech system by addressing it with a name chosen in advance, in this case the expression "wake up" (line 1), to which the system responds with an audible sound to signal it is listening (line 2). Similar to instance 2, what follows is a sequence of turns in which the speech system asks the participant to select an alternative phone type to the one initially requested (lines 4, 8 and 10). However, the participant rejects the two phone types proposed by the system (line 7, "Neither"), triggering a series of misrecognitions as the system is expecting a phone type selection from the proposed options instead of a rejection. For the first misunderstanding, the system requests a directive repetition ("Pardon?", in line 8). The second time around, it adds a short apology marker "sorry" in frontal position, followed by an explanation of the problem and another request for repetition (line 10, "Sorry, I didn't understand. Please try again"). So far, although more elaborate, this system behaviour is similar to the one designed for Study 1. That is, the system relies on short apologies to assure participants they are not to blame for the misunderstanding, and to encourage them to try again.

However, in line 13, the apology introduces the illocutionary act of regret to account for the multiple failures and further elaborates on the notion of being at fault for the communication breakdown (line 13, "I'm so sorry, I still didn't understand. Clearly I'm limited"), inviting the user to repeat the selection ("Please try one more time."). The apology starts with a slightly more upgraded expression of regret ("I'm so sorry"), followed by an account for the ongoing inability to proceed with the task ("I still didn't understand") and an explicit account of the system's responsibility for the fault ("Clearly I'm limited"). This last apologetic turn includes multiple strategic elements, to which participants orient as a marked system response. As such, several participants refer to this turn in the subsequent interviews, as we later discuss.

Instance 4 below shows a similar progression. Here, the apology is triggered after the system mistakenly tries to confirm a song title (line 4, "Colder Weather. Correct?"). As a result, the participant corrects the speech system with the rejection "no" followed by the repetition of the desired song title (line 5, "No. Come Together"). However, due to technical limitations, the system can only understand the rejection "no", but misses the desired song title.

Instance 6
Participant 7 – 1:04:15
1 P: Garbanzo
2 S: (audible beep)
3 P: Play Beatles song called Come Together.
4 S: Colder Weather. Correct?
5 P: No. Come Together.
6 S: Oh! I'm sorry. So embarrass[sing. [What would you like?
7 P: [aww..]
8 P: The Beatles singing Come Together.
9 S: Great! I found more than one copy for Come Together. I'll play them
10   all.

In line 6, the system responds with an elaborate apology containing an expression of negative surprise and change-of-state token "oh!" in initial position (Heritage, 1984), indicating the user rejection was unexpected. This is then followed by the apology "I'm sorry". The strong sense of regret is emphasized by an emotional disclosure "so embarrassing", which causes the participant to mumble "aww" (line 7). The participant clearly orients to this phrase as an expression of human emotion, which is later verbalized during the interview as evoking an undefined negative feeling on the part of the participant, broadly characterized as an inappropriate reaction when coming from a machine.

The highly elaborate and recursive apologetic sequences suggest that the system seeks not only some absolution for the communication issue, but also the user's affiliation. By over-emphasizing responsibility and alluding to human emotions, apologizers also portray a negative judgmental stance toward themselves and the action undertaken. This, in turn, would promote a recipient's response similar to the one triggered by self-critiques and expressions of self-contempt in human-human communication. That is, the recipient of the elaborate apology disagrees with the apologizer and thus, supports the speaker, reassuring that their "morality" is preserved and publicly acknowledged instead (Pomerantz, 1984), which here would break the suspension of disbelief requisite for human-machine interaction.

The quality of such mixed emotional reactions toward the elaborated apologies become more evident in the interview data. In what follows, we incorporate some excerpts from interviews to further illustrate this evaluation.

7.9 "I do like polite, but I don't need apologetic"

These excerpts indicate that participants valued brief apologies along the lines of a politeness discourse marker rather than an attempt at an apology per se. As this study evidences, an over-apologetic tone breaks the suspense of disbelief, creating a push back in terms of a negative threatening face act. As the system imposes a regretful tone on the interaction and makes the user feel either guilty for triggering an apology, or awkward for interacting with a system that evokes this type of human emotion.

1. "Hmm the thing where it said (.) the first thing was 'I'll try not to mess this up' couple of calls ago, or (.) 'I'm so embarrassed,' to me *it sounded a little neurotic* (...) "
2. "Cut off the apologies (.) I mean (.) it's a computer, why does it have to apologize?"
3. "I'd get rid of the sorry word...I do like polite, but I don't need apologetic"

Interestingly, although participants in general judged elaborate apologies as a negative trait, for participants who prefer a more interactive conversational style with the system, this evaluation was slightly different from those who preferred a more effi-cient interaction. The conceptual distinction between predominantly efficient versus predominantly interactive users in spoken human-machine communication was in-troduced in Winter, Shmueli & Grost (2013) and Carbaugh et. al. (2016). While the pre-dominantly interactive users tended to feel guilty or awkward for having triggered such system responses, the efficiency biased users mainly assessed the over-apolo-getic tone as unnecessary coming from a computer and thus, simply a waste of time. Note that in each case, the semantic structure of the apology, its conversational posi-tion, or its verbiage are not problematized, rather the fundamental idea that the sys-tem purport to have human emotion is rendered unacceptable.

Such apologetic expressions also made the prompts lengthier than desired, add-ing time toward the goal completion as apologies got gradually more elaborated with each system turn. While for some participants acknowledging impact and limitations of one's actions may be perceived as polite behaviour, efficient participants perceived it as adding to their frustration exactly because they were continuously attempting a given command or task. Lengthy apologies backfired for these users in particular, rendering the system not only as incompetent but also as inefficient. The progression, however, was positively perceived in terms of acknowledging the severity of the issue.

This suggests that the status of personhood granted to the system is variable on grounds of cultural premises held by users that informs whether users are willing to suspend their disbelief and enter into a human-like interaction with the system, ver-sus those who do not grant the system this kind of personhood and prefer it behave as an efficiently functional machine. As suggested by Carbaugh (2007), cultural prem-ises often implicate other radiants of implicit meaning and so the granting of a tem-porary or contingent personhood to a machine may be tied to the consequences of such a thing given, i.e. implications of relationship development, concern for the feel-ings and wants of others, etc. that may be connected to that attribution.

While no user appears to grant the system full human interactional status, as ev-idenced by users resistance to responding to the system's emotional apologies with supportive affirmations, it is clear that some variation exists between those who dis-like the more elaborate emotional apologies on the grounds that there is something fundamentally wrong with a machine trying to be *that* human, versus those who dis-like it because machines are *a priori* always already not human making the elaborate apologies a waste of time. Some explanation for this may lie in user premises of emo-tion, where users who prefer a predominantly elaborate interactional style are willing to grant enough personhood to the system to accept some mimicry of human dimen-sions of interaction, but where the system's purported experience of emotion is too far. In the case of some predominantly efficient users, very little personhood is granted to the machine and most attempts at human interactional mimicry beyond the minimum necessary for task completion are evaluated negatively. The question then is where and in what domain of human interaction do users draw the line? In this case, a premise of personhood and emotion might be articulated that:

1. *machines are not granted human personhood,*

with an accompanying normative rule that:

2. *machines, even when interactionally capable of doing so, ought not purport to have human emotions.*

For predominantly interactive users however:

3. *machines ought to some degree play at being human.*

While for predominantly efficient users:

4. *machines ought not play at being human beyond the minimum interactional capacity to accomplish the task.*

7.10   Discussion and Design Considerations

Analysing different user orientations toward apologies, we argue that not all relational acts are created equal, in part because users' interpretations of the goodness of a particular speech act from the system relies on cultural premises for personhood, social relations, emotion, and communication where some acts may be presumed as appropriate relationally generative interactions performed with a proper interactional partner (a humanistic car), or as inappropriate intimations misusing particular communicative acts with the wrong kind of interactional partner.

In general, our observations indicate that elaborated apologies were negatively evaluated by most participants, and that the reasons behind this evaluation can be attributed to two different rationales. For some participants, the issue stemmed from feeling guilty because of their unintended contribution to the system's failure or violation of rules, or from perceiving the system as awkward and 'too neurotic'. However, for others, the in-car system was characterized as a mere tool and as such, displays of human-like emotions in length or intensity were seen as an unnecessary waste of time. A key observation is that participants displayed interactional style seems to play a central role in terms of how they evaluated elaborated apologies: evoking a sense of awkwardness on the one hand, or an irritating waste of time on the other. This observation correlates with previous findings stating that there are competing cultural norms in participants' communication about the appropriate relationship between machines and humans (Molina-Markham et. al., 2016).). Furthermore, the authors explain that participants' interaction with the system indicate a lack of crystallization of what constitutes proper interaction with machines and appropriate system responses.

Generally speaking, apologies seem to be related to self-presentational goals and the judgements people make about the transgressor. In our study, we observed that the use of apologies has an affect on the judgements related to the identity of the system. Similar to previous research, we found that the desired user evaluation was triggered by one apology strategy, rather than by the use of multiple strategies. It has been stated that speaker responsibility tends to be implied by each of the apology strategies (Scher, 1989). If explicit, these elements may evoke a sense of recipient's relational obligation, which was evidenced during interviews as participants shared their perceptions of the system as having a 'neurotic' persona, or evaluate the exchange as a 'waste of time' or perceiving 'insincerity' in the system responses. Given the implicational nature of apology strategies, some of these strategies may remain implied and still have effect on the judgement of the transgressor (Blum-Kulka, 1997). Heritage and Raymond (2016) propose that different formats of apologizing (e.g., bare vs. expanded) may appear proportionally to certain dimensions of the violation, such as whether it is minimal and proximate to the participants, or distal offenses that require more elaboration from the transgressor. Our findings seem to orient to these prior observations, as participants' reactions to concise versus elaborate apologies further strengthen the notion of a degree of appropriateness for apology elaborations: brief apologies are valuable for grounding during communication breakdowns, as politeness markers, as well as to indicate source of error. However, 'genuine' elaborate apologies that evoke emotion are negatively evaluated, as the type of issues encountered during exchanges with the speech system are generally not perceived as requiring such extensive interactional work.

This can in part be explained using Brown and Levinson's (1987) politeness theory, which suggests that certain acts, such as apologies, are inherently threatening to the face wants of both speaker and hearer. As a result, interlocutors may choose to "pay face" as a way of advertising an upcoming threat to face, as in the prefatory work done to assure you have conveyed that you understand and appreciate the importance of someone's time before asking for a ride to the airport. In this way, instances of upgraded apology, as in the case of the system's use of "so sorry" rather than "sorry" may work to advertise, or seek to compensate for, a face threatening act that is not evaluated by the user as proportionally appropriate to the advertisement of it.

Needless to say, as speech technologies continue to mature and spread across everyday environments such as the vehicle, design decisions to build a well-rounded voice assistant become even more salient. Users' expectations are invariably related to the social aspect of interacting with technology through speech. As such, appropriate levels of sociability in the voice assistant, which align well with cultural premises about personhood, relations, emotion and communication, seem to influence perceptions of trust, usefulness, and overall enjoyment. As the user-voice assistant relationship evolves, a key design consideration—and an ongoing challenge— is the degree to which speech systems adapt to users' changing expectations for cultural and communicative appropriateness.

References

Akgun, M., Cagiltay, K., and Zeyrek, D. 2010. The effect of apologetic error messages and mood states on computer users' self-appraisal of performance. Journal of Pragmatics, 42 (9):2430-2448.

Blum-Kulka, S., and Olshtain, E. 1984. Requests and apologies: A cross-cultural study of speech act realization patterns CCSARP. Applied linguistics, 5:196.

Blum-Kulka, S. 1997. Dinner talk: Cultural patterns of sociability and socialization in family discourse. Lawrence Erlbaum Associates. Mahwah (NJ)

Brown, P., Levinson, S. C. 1987. Politeness: Some universals in language usage. Cambridge University Press. Cambridge.

Carbaugh, D. 2005. Cultures in conversation. Lawrence Erlbaum. (NJ)

Carbaugh, D. 2007. Cultural discourse analysis: Communication practices and intercultural encounters. Journal of Intercultural Communication Research 36:167–182. doi:10.1080/17475750701737090

Carbaugh, D. 2012. A communication theory of culture. In: (A. Kurylo, ed) Inter/Cultural Communication: Representation and Construction of Culture, Sage. Thousand Oaks, pp. 69-87.

Carbaugh, D., Winter, U., Molina-Markham, E., Van Over, B., Lie, S. and Grost, T. 2016. A Model for Investigating Cultural Dimensions of Communication in the Car. In: Theoretical Issues in Ergonomics Science 17(3):304-323.

Fortin, D. and Dholakia, R. 2005. Interactivity and vividness effects on social presence and involvement with a web-based advertisement. Journal of Business Research 58:387-396

Fraser, B. 1981. On apologizing. In: (F. Coulmas, ed) Conversational Routine: Explorations In Standardized Communication Situations and Prepatterned Speech. pp. 259-272.

Gefen, D. and Straub, D. 2003. Consumer trust in B2C e-Commerce and the importance of social presence: Experiments in e-Products and e-Services. Omega 32(6):407-424.

Gefen, D. and Straub, D. 2004. Managing user trust in B2C e-Services. E-Service Journal. Indiana University Press.

Goffman, E. 1959. The presentation of self in everyday life. Doubleday. Garden City (NY)

Gumperz, J. J. 1982. Discourse strategies. Cambridge University Press. Cambridge.

Gumperz, J. J. 1992. Contextualization and understanding. In: (A. Duranti and C. Goodwin, eds) Rethinking context: Language as an interactive phenomenon. Cambridge University Press. Cambridge, pp. 229-252.

Hassanein, K. and Head, M.M. 2004. Manipulating social presence through the web interface and its impact on consumer attitude towards online shopping. In: McMaster eBusiness Research Centre (MeRC) Working Paper Series, WP #10. Hamilton, Ontario.

Hassanein, K. and Head, M. 2007. Manipulating perceived social presence through the web interface and its impact on attitude toward online shopping. International J. of Human-Computer Studies 65:689-708

Heritage, J. and Raymond, C. 2016. Are explicit apologies proportional to the offenses they address? Discourse Processes 53(1-2):5-25.

Hymes, D. 1972. Models for the interaction of language and social life. In: (J. J. Gumperz and D. Hymes, eds) Directions in sociolinguistics: The ethnography of communication. Blackwell. New York, pp.35–71.

Karahanna, E. and Straub, D. 1999. The psychological origins of perceived usefulness and ease-of-use. Information and Management 35:237-250.

Kumar, N. and Benbasat, I. 2006. The influence of recommendations and consumer reviews on evaluations of websites. Information System Research 17(4):425-439.

Levinson, S. 1997. Contextualising contextualisation cues. In: (S. Eerdmans, C. Prevignano and P. Thibault, eds) Discussing Communication Analysis 1: John Gumperz. Beta Press. Lausanne, pp. 24-30.

Malinowski, B. 1923. The problem of meaning in primitive languages. In: (C. K. Ogden, ed) The meaning of meaning. HBJ. Orlando

McLaughlin, M. L. and Cody, M. J. 1982. Awkward silences: Behavioral antecedents and consequences of the conversational lapse. Human Communication Research, 8(4), 299-316.

Molina-Markham, E., van Over, B., Lie, S. and D. Carbaugh. 2016. "You can do it baby": Cultural norms of directive sequences with an in-car speech system. Communication Quarterly 64(3):324-347.

Nass, C. and Brave, S. 2005. Wired for speech. How voice activates and advances the human-computer relationship, MIT Press.

Nielsen, J. 1995. Usability engineering. Morgan Kaufmann Publishers Inc. San Francisco (CA)

Norman, D. and S. Draper. 1986. User centered system design; New perspectives on Human-Computer Interaction. Lawrence Erlbaum. Mahwah (NJ)

Nowak, K. L. and Biocca, F. 2003. The effect of the agency and anthropomorphism on users' sense of telepresence, copresence, and social presence in virtual environments. Presence: Teleoperators and Virtual environments, 12(5):481-494.

Olshtain, E. and Cohen, A. 1983. Apology: A speech act set. Sociolinguistics and language acquisition, 18-35.

Park, E.K., Lee, K.M., and Shin, D.H. 2015. Social responses to conversational TV VUI: Apology and voice. International Journal of Technology and Human Interaction (IJTHI), 11(1), 17-32.

Pearl, C. 2016. Designing voice user interfaces: Principles of conversational experiences. O'Reilly Media.

Pomerantz, A. 1984. Agreeing and disagreeing with assessments: Some features of preferred/dispreferred turn shapes. In: (J.M. Atkinson and J. Heritage, eds) Structures of Social Action. Cambridge University Press. Cambridge, pp. 57-101.

Poutiainen, S. 2015. Interpersonal ideology. The International Encyclopedia of Language and Social Interaction. 1–6.

Rosenbaun, L., Rafaeli, S. and Kurzon, D. 2016. Blurring the boundaries between domestic and digital spheres: Competing engagements in public Google Hangouts. Pragmatics 26(2): 291-314.

Scher, S. J., and Darley, J. M. 1997. How effective are the things people say to apologize? Effects of the realization of the apology speech act. Journal of Psycholinguistic Research, 26(1):127-140.

Scher, S. J., and Huff, C. W. 1991. Apologies and causes of transgressions: Further examination of the role of identity in the remedial process. In: Meetings of the Midwestern Psychological Association, Chicago (IL)

Scollo, M. 2011. Cultural approaches to discourse analysis: A theoretical and methodological conversation with special focus on Donal Carbaugh's Cultural Discourse Theory. Journal of Multicultural Discourses, 6: 1–32. doi:10.1080/17447143.2010.536550

Serenko, A. 2007. Are interface agents scapegoats? Attributions of responsibility in human–agent interaction. Interacting with computers, 19(2):293-303.

Skarlicki, D. P., Folger, R., and Gee, J. 2004. When social accounts backfire: The exacerbating effects of a polite message or an apology on reactions to an unfair outcome. Journal of Applied Social Psychology, 34(2):322-341.

Vredenburg, K., and M. Butler. 1996. Current practice and future directions in user-centered design. In: Proceedings of Usability Professionals' Association Fifth Annual Conference, Copper Mountain (CO)

Winter, U., Shmueli, Y., and T. Grost. 2013. Interaction styles in use of automotive interfaces. In: Proceedings of the Afeka AVIOS 2013 Speech Processing Conference, Tel Aviv, Israel.

8 "Ok, talk to you later": Practices of Ending and Switching Tasks in Interactions with an In-Car Voice Enabled Interface

8.1 Introduction

Individuals' expectations for the ending of interactions vary culturally and impact interpretations of and satisfaction with an encounter. How voice activated technology ends interactions is thus an important area of concern for system designers. In this chapter, we examine communication between an in-car speech enabled computer system and drivers, focusing on the way in which system tasks—such as the making of a phone call or playing of a song—come to a close. We analyse instances when task endings do not appear to proceed as participants anticipate in order to examine a possible cultural norm informing communication in the car. This research expands traditional analyses in terms of user preferences and personas by focusing on actual user practices in interaction.

  While speaking with an in-car speech activated system during a research study, four participants laughed in response to the system's ending an interaction with the following phrase, "OK, talk to you later." One example is included below:

Instance 1
Participant 5 - 1:07:35
P = Participant, S = System
1 P: ((parks car and scrolls through contacts with touch)) OK. I can
2   scroll to touch, right. Anyway. Home.
3 S: OK, talk to you later.
4 P: ((laughs)) Hey that's kind of cute.

Participants' laughter in these instances could indicate that the in-car speech system's response was in some ways unanticipated (Coupland and Coupland 2001). Participants likely did not expect the system to reply at this point in the conversation with an acknowledgement that it would interact with them at a future time. Their expectations for how the encounter would end may not have been met. Researchers argue that people's expectations for the unfolding of an interaction are culturally shaped (Hymes, 1972; Philipsen, 2002; Carbaugh, 2005, 2007). Norms of interaction guide what participants deem appropriate at each point in an encounter from the beginning to the end (Scollon and Scollon, 1981). While in this particular case the use of the phrase, "OK, talk to you later," appears to have been at least to some extent appreciated by participants—as when Participant 5 notes "that's kind of cute"—there is also the possibility, which we will elaborate further below, of an

https://doi.org/10.1515/9783110519006-008

interactant ending an encounter in an unanticipated way that is considered abrupt, rude, or inappropriate by a conversational partner. This failed ending may negatively influence a person's view of the other person (or the in-car system) with whom they are speaking, and it may even taint possible future encounters.

In designing a voice activated system, system designers draw on assumptions about the nature of the conversants with whom their system will interact and those users' expectations regarding dialogue flow. Given that premises regarding the nature of acting in this type of situation may vary between cultural communities, it is important for designers to take into account how their own premises could influence design in a way that may not be pleasing to users. As Murray (2012) writes "the design of digital objects is a cultural practice like writing a book or making a film" (p. 1). She further explains, "In order to make truly intuitive interfaces, designers must be hyperaware of the conventions by which we make sense of the world— conventions that govern our navigation of space, our use of tools, and our engagement with media" (Murray, 2012, p. 10). Therefore, it is essential to consider cultural premises when designing a digital system.

In this chapter, we analyse the ending of tasks in communication between a driver and an in-car speech enabled system. We present excerpts from recordings of these interactions from western Massachusetts. We adopt a cultural discourse theory (Berry, 2009; Carbaugh, 1988, 2007, 2012; Scollo, 2011) perspective in order to identify assumptions about the communication situation that seem to be made by both the participants as well as by system designers. Finally, we compare these task ending practices of participants from western Massachusetts with task ending practices of participants from data collected in mainland China.

Instances demonstrate that drivers in the United States used both a voice command, such as "end radio" or "stop music," and the touch screen to end a task. There were also some participants who said "back" or "home" when they wanted to end a task, and many participants who never ended a task at all, but rather switched to another task without ending the first. As the analysis below demonstrates, these task endings suggest that participants are coming to the interaction with expectations regarding the ability of the system to listen continuously to them. Some participants appear to view the communication event with the system as part of a longer communication situation that does not necessarily contain distinct ending points. These assumptions have design implications since there could be adjustments made to accommodate user preference in task ending and user perception of situation boundaries.

8.2 Theoretical Background

Previous scholarship has explored the dynamic of how participants end interactions of varying structure and formality in different contexts and cultures and at different points in time (Goffman, 1972, 1974; Hartford & Bardovi-Harlig, 1992; Knapp et al., 1973; Schegloff & Sacks, 1973; van Over, 2014). In achieving closings, participants transition from one communication event or situation to another. In their frequently cited piece on the topic, Schegloff and Sacks (1973) assert that closings are oriented to by conversationalists as having "a proper character" (p. 320); they identify the "closing section" of a conversation as initiated by a "pre-closing" and containing at minimum a terminal exchange, although it may contain other component parts (p. 317). Goffman (1974) notes that closings are connected to participants' framing of an interaction as guided by certain "principles of organization which govern events" (p. 10). Thus, the characteristics of endings depend on what genre of event partici- pants deem a certain encounter to be; for example, bringing to a close an introduc- tory meeting between new colleagues in Finland (Carbaugh, 2005) will be done differently than participating in silence at the end of a meeting for business among Quakers (Molina-Markham, 2014) or "dissolving" a town meeting in New England (Townsend, 2009).

Communication practices of closing are also closely connected to the nature of participants' relationships and the development of those relationships. Goffman (1971) characterizes both greetings and farewells as "access rituals," which he de- fines as "ritual displays that mark a change in degree of access" (p. 79). How this access is negotiated has important consequences for a relationship. Goffman (1971) explains, "the goodbye brings the encounter to an unambiguous close, sums up the consequence of the encounter for the relationship, and bolsters the relationship for the anticipated period of no contact" (p. 79). The importance of leave-taking in re- flecting and informing relationships means that, for example, how one takes leave from a social gathering is dependent on the bonds understood to exist between host and guest (Fitch, 1990), and the parting of those considered social equals is likely to be different from those who are viewed as having varying degrees of power or con- trol (DuFon, 2010). In many cases, strategies for ensuring good relations are em- ployed, such as launching a new topic in a way that demonstrates other attentive- ness (Bolden, 2008) or blaming an external factor in a pre-closing in order to provide a legitimate reason for why an interaction must end and not imply that one wants to end it (Takami, 2002). Bolden (2008) notes, "the collaborative nature of leave-taking is an important resource for maintaining and reaffirming interpersonal relationships" (p. 101).

Misunderstanding can occur when participants from different speech communi- ties have different expectations for how an interaction should end. Practices of leave-taking vary widely within and across cultures; for example, Omar (1993) finds that the order of features in Kiswahili closings is less strict than in English closings, and Dogancay (1990) observes that among English speakers, both participants in an interaction may end by saying "goodbye," while among members of a Turkish speech community, there is one phrase used by the person who is leaving, which translates to "I recommend you to God," and a different phrase used in response by the person who is staying, which translates as "laughingly" (p. 60). Scollon and Scollon (1981) explore possible misunderstanding in their examination of Athabaskan-English communication. They explain that while an American or Canadian English speaker may be inclined to end a conversation with an indication that he or she hopes to continue talking with the other person in the future, an Athabaskan, who has been socialized in different communication patterns, would feel it was bad luck to make predictions about the future. Thus, an expression such as "talk to you later," as used by the in-car speech system that we studied here, would be avoided by an Athabaskan. Scollon and Scollon (1981) describe how this avoidance could become problematic, writing, "The Athabaskan, being careful of courting bad luck, may quite unknowingly signal to the English speaker the worst possibility, that there is no hope of getting together again to speak." (p. 27). In this way, the ending of an interaction has consequences for how people view the success of the encounter and the possibility for future positive interactions.

This chapter expands on prior research by examining practices of ending interactions with a new type of interlocutor—an in-car speech activated computer system. While prior research has looked at how people speak with machines (Friedman, 1997; Nass & Brave, 2005; Nass & Yen, 2010; Turkle, 2011) and also at how people interact with each other in the car (Laurier, Brown, & Lorimer, 2007; Laurier et al., 2008; Haddington, 2010), speaking with a computerized speech system in the car has not been as deeply explored in the research literature. In this way, this analysis builds on past understandings of the closing of communication events by looking at a new context. The analysis thus reflects McKay's (2013) notion that "a user interface is essentially a conversation between users and a product to perform tasks that achieve users' goals" (p. 3). McKay (2013) asserts that this context is distinctive because the "conversation" is conducted through "the language of UI instead of natural language" (p. 3). This analysis is of use to system designers because it goes beyond describing characteristics of typified users, and instead, focuses on concrete practices of acting in the car. In other words, we highlight concrete ways of enacting a social role through discourse and interaction in a particular cultural context (Geertz, 1972). Closings are important to consider in designing speech-activated systems because they can influence participants' opinions about the system and willingness to interact further with the system in the future.

In this chapter, we analyse participants' preferences for ending interactions with the system. This analysis will include a summary of the variety of ways that participants chose to end tasks, as well as a closer look at some instances of task endings in which participant expectations and system actions appear not to have been in alignment. Examination of these instances adds depth and insight into the previously identified preferences. We conclude with a discussion of the implications of these findings, which includes the formulation of a cultural premise that we propose serves as a foundation for, and is informed by, participants' communication practices in this context.

8.3 Analysis

Although we initially conducted driving sessions with 26 participants, we have chosen not to directly include the first 6 participants in the analysis below because the ability to switch between tasks was not fully operational for these participants, due to system development delays. (It should be noted that the data from the first 6 participants does inform this analysis, however, as indicated by the inclusion of the excerpt from Participant 5's driving session in the introduction.) Participants initiated a verbal interaction with the system by touching the microphone button on the touch screen. The system would then produce a chime sound, or "ding," and the participant would tell the system what he or she wanted it to do. When they wanted the system to stop doing something, participants had the option to end the radio or end the music by either touch or by voice, but they were only able to end phone calls by touch. Ending a task by touch was done by simply pushing the "end" button on the tablet screen. In order to end a task by voice, it was necessary to first touch the microphone button and then give a verbal command to end the task. Phone calls could only be ended by touch because of technological restrictions. During a phone call, it would be difficult for the system to know whether the driver was talking to the call recipient or giving a command to the system. There is also the possibility that the driver would touch the microphone button to give a command during a phone call, but if the call recipient did not realize that the driver had done this, he or she might speak and confuse the system because it might interpret what the call recipient had said as a command. As a result of this possible confusion, the system was designed so that phone calls could only be ended by touch.

8.3.1 Preferences for ending tasks

In reviewing the recordings of the 20 participants, we counted 49 instances in which participants ended a task by voice and 135 instances in which a participant ended a task by touch. One example of each is given below: Participant 24 ends the music by using a voice command, and Participant 9 ends a phone call by touching the "end" button.

Instance 2
Participant 24 - 31:01
1   ((music playing))
2 P: ((touches microphone button and system dings))
3 S: What kind of music would you like to hear?=
4 P: No music. Stop music.
5 S: ((music stops playing and system returns to home screen))

Instance 3

Participant 9 - 24:27

C = Callee

1   P: OK, see ya.

2   C: (Bye).

3   P: ((touches end button and system returns to home screen))

Eight of these voice and touch instances overlap, in that the participant used both voice and touch to end the task—such as when Participant 8 said "end" while touching the "end" button on the screen.

Instance 4

Participant 8 - 24:51

1   C: Enjoyed talking to you and uh good luck I hope that th- the results

2   of your experiment continue now that you figured that thing out.

3   P: Yup thanks.

4   C: OK. Call me again.

5   P: Alright. Bye Mom=

6   C: Bye. Love you.

7   P: End. ((touches end button as speaks and system returns to home

8   screen))

9   P: End call.

These numbers indicate a preference for ending tasks by touch. However, as mentioned, participants were only permitted to end phone calls by touch, which could have inflated the number of touch endings because there was no option in these cases. We found that of the 135 instances in which a participant ended a task by touch, in 64 instances, the participant was ending a phone call. In order to account for the requirement to only use a touch command to end a phone call, we could discount these 64 cases, and only consider the 71 other touch endings (of the music or radio), in which the participant had a choice of touch or voice. It is likely, however, that at least some of the 64 phone call endings would have been touch endings even if there had been an option to use voice, so the number of touch endings would likely be greater than 71. Thus, 71 (or more) touch endings versus 49 voice endings still seems to indicate a preference for ending by touch. The preference for ending by touch may be due to the fact that ending by touch only required one step—that of touching the screen—while ending by voice required two steps: touching the microphone button on the screen and then giving a verbal command.

Although the system had multimodal capabilities, participants were instructed at the beginning of the session to only use voice commands with the system while they were driving—except when they needed to touch the microphone button to start a voice interaction or to touch the end button to end a task. Thus, although participants were permitted to end tasks through touch, one might reasonably expect a pattern of dispreference for this mode to emerge as they had been told not to use touch for all other tasks. However, the high number of touch endings indicates that no such dispreference was active.

We also counted approximately 91 instances of participants switching tasks without ending tasks. These switches included changes between radio stations, between listening to the radio or music and making a phone call, and between artists, albums, or songs (we chose not to include going to the next song on an album as a task switch since this switch is not a change to a different task). We have included an instance of a switch between radio stations below. In this case, Participant 20 does not end the sports channel that he is listening to before going to a different radio station.

Instance 5
Participant 20 - 21:34
1   ((radio playing))
2   P: ((touches microphone button and system dings))
3   S: What sports channel do you want to hear?
4   P: Go to NPR.
5   (5)
6   S: Hold on. (5) Tuning radio to 88.5 FM WFCR.
7   ((radio station changes))

When compared with the 135 touch endings and 49 voice endings, there would seem to be a preference first for touch endings, next for switching tasks, and finally for giving a voice command to end a task. However, if the 64 instances of ending phone calls are discounted because they required the participant to use a touch ending, as discussed above, then there is slightly more of a preference for switching tasks (91) rather than for touch endings (approximately 71). It is, however, difficult to know if participants were simply switching tasks in order to experiment with the system, since they had been told during the introduction to the study that they were helping the researchers to test a prototype. They were also sometimes introduced to the switching functionality during the mid-session interview by the interviewer if they did not discover it on their own during the first part of the drive, which could have increased their likelihood to use this functionality. It is possible that had participants been alone in the car they might not have felt the need to continue to test the system by switching between tasks, but would have simply ended the current task. Therefore, we will now consider in more depth several instances of task endings that appear to have been somewhat problematic for users in order to explore what these instances might reveal about users' task ending practices and preferences.

8.3.2 Further consideration of instances of task endings

We will now consider several instances of endings in which there appears to be misalignment between participants' expectations and the system's responses. We will suggest that these misalignments are related to premises of ending interactions.

8.3.2.1 Touching the microphone button
Participants sometimes forgot to touch the microphone button before giving a voice command to end a task, such as Participant 24 below.

Instance 6
Participant 24 - 23:13
1   ((radio playing))
2   P: End radio. (3) ((touches end radio button, radio ends, and system
3   returns to home screen))
4   P: Phone.

Beginning to speak before touching the microphone button occurred frequently throughout the data, not just during task endings. In addition to forgetting to touch the button before ending a task, participants also forgot to touch the button at the beginning of tasks. The example from Participant 24 above was one of at least five instances in which Participant 24 forgot to touch the microphone button before giving a verbal command to either initiate or end a task. Toward the end of her driving session (around fifty-two minutes into the session), Participant 24 forgot to touch the microphone button before giving a verbal command to play the radio, and after this final instance, she reflected aloud:

> This reminds me of Star Trek when they- I'm a really big Star Trek: The Next Generation fan. Sometimes they like hit their communicators to talk, and then sometimes they're just like 'Computer' whatever, like they just talk to like the whole ship, and they're not even like- they're not even hitting their communicator. And it always bothered me cause I always thought, how does the computer know the difference between their conversation in the ship, and like- ((laughs)) You know they should be like hitting the button every time. That's what I feel like when I have to keep hitting this button here. I keep forgetting to do it. Cause I keep thinking it's just going to read my mind. ((laughs))

Thus, for some participants at least, it seemed natural to simply begin speaking to the system without a need to initiate the interaction by touching a button. Participant 17 noted during the mid-session interview when asked what he thought of the system so far, "It seems kind of strange that you have to keep touching the microphone button. I think it should either be engaged or- once you engage it, it should stay that way until you turn it off, I would think." This beginning to speak without initiating the dialog through touch suggests a view of the system as always on or continuously listening—an expectation which is not met by the current system design.

8.3.2.2 Trouble during task switching

Some participants did not seem to realize at first that switching was possible or how to do it. For example, Participant 17 had difficulty figuring out how to accomplish switching. In the instance below he tries to switch XM channels, but since he first says "end radio," the system shuts off before responding to his request for "seven." The system is designed to hear only a statement about desired action, at which point it closes the sequence, while this user's turn design suggests a preference for an ongoing conversation where multiple tasks can be accomplished in sequence moving from ending the current act, to establishing a new act in the same interaction.

Instance 7

Participant 17 - 1:01:14
1 P: ((touches microphone button and system dings)) End radio. Seven.
2 S: ((radio ends and system returns to home screen))

This difficulty in switching happened twice during Participant 17's driving session. As a result, Participant 17 did not switch tasks at all during his session. His attempts at switching became voice endings.

The desire to switch tasks (although unsuccessful) aligns with the tendency to switch that is evident in the 91 task switches that other participants did accomplish successfully. Participants' switching reveals an understanding of the communication situation as ongoing, rather than one marked by distinct ending points. Also, since Participant 17's attempts to switch resulted in voice endings, these instances suggest that in some cases when a participant uses voice to end a task, he or she is in fact trying to switch tasks rather than end a task.

8.3.2.3 Going "back"

Several participants used the term "back," when they wanted to return to a previous screen and switch tasks. However, the system's response to the command "back" was to repeat the previous action. In the instance below, Participant 23 attempts to use the commands "back" and "home" in order to get to a previous screen before giving a directive for a different task. The system's response of repeating the previous action (lines 4-5) appears to confuse him, as indicated by line 6 when he says, "hmm."

Instance 8

Participant 23 - 21:09

1   P: ((touches microphone button and system dings))
2   S: What music station do you want to hear?
3   P: Um (1.5) Go back.
4   S: Tuning radio to one hundred and four point one FM radio one
5   hundred and four point one WMRQ.
6   P: Hmm ((touches microphone button)) (3) Home.
7   S: Could you repeat that please?
8   P: Home.
9   S: OK, talk to you later. ((participant laughs))

Below, Participant 7 attempts to use "go back" to return to the screen that al-lowed her to choose genres of stations so that she could switch tasks. However, the system responds in line 8 by repeating its previous action from line 3, when it told her that it could not find the artist that she had requested. Participant 7 does not seem to understand that the system has repeated its previous action, and she presses the back button several times in a continued attempt to return to a previous screen.

Instance 9

Participant 7 - 42:32

1   S: Please review your specified artist once more.
2   P: ((unclear))
3   S: I cannot find this artist in your collection. I am sorry.
4   P: OK. Um. Let's see. How about, um, something simple. (1.5) Um=
5   S: Pardon?
6   P: Mm. (4) Go back to g- Uh, can you go back? (3) Can you go back?
7   ((touches back button)) Go back.
8   S: I cannot find this artist in your collection. I am [sorry.]
9   P: [OK.] OK. OK=
10 S: Please review your specified artist once more.
11 P: ((touches back button again twice))

On line 5 the system is waiting to hear a recognizable command and apparently hears something on line 6 it understands as the "back" command and takes this action. In doing so, it has returned to the previous user-initiated action, a request to play an artist. Because the participant continues to say "Go back" after the system has re-initiated its prior action, listening for an artist name, "Go back" is likely heard by the system as an artist name that the system cannot recognize. We suggest that the system has interpreted "Go back" in this way based on the system's next turn of notifying the user that this artist cannot be found. At this point, the user responds "OK. OK. OK." perhaps expressing agitation with the interaction, which the system hears as yet another attempt to offer an artist name and responds yet again that this artist cannot be found.

Because the user anticipates that the "back" command will return them to a point further back in the interaction, while the system treats "back" as a return to the user's last turn, a misalignment of expectations for what "back" should accomplish is manifested. This is suggestive of a shift in understanding about when "this" began: a single turn ago, or an event ago. A conflict appears to exist here between machine organization and the organization of conversation between people. While a person interprets interaction in terms of a turn-based sequential organization (Sacks, Schegloff, and Jefferson, 1974), a machine functions differently, unless told to do otherwise. A user can input a directive (one turn from the user's view) and the system then responds (another turn from the user's view, making a pair), but from the system's view, in the space of that turn it has taken many micro-actions and executed many internal commands, to bring about the result the user directed. So when a machine "goes back" it goes back to the immediate prior internal command "tuning radio" but when a user says "go back" he or she means go back to the prior pair-part, or two conversational turns.

These examples of different interpretations of the command "back" on the part of the system versus participants again indicate a desire to be able to switch tasks without ending previous tasks. The participant would like to return to an earlier step in the interaction. The interaction is thus seen as part of a larger situation, rather than a distinct and separate event.

8.4 Discussion

The frequent use of switching by participants, in addition to certain forms of misalignment in endings, appear to indicate an understanding of the interaction with the in-car system as an ongoing situation that does not begin again completely when the participant touches the microphone button. Misalignments, such as forgetting to press the microphone button before ending a task verbally, trouble switching tasks, and the use of the command "back," reveal a cultural norm that could be formulated as: *when interacting with a system participant (i.e., the in-car system), if this is done properly, one should be able to switch to a new task without explicitly ending a previous task.*

As Participant 11 explained during the mid-session interview, when the interviewer commented that she had a tendency to go from one task directly to another without ending the first and asked if this practice would be her preference when interacting with the S: "Yeah, I think it should be open and free. Um, so that people don't have to think about, 'Oh, well I want to make a phone call. I got to end the music and then-' Yeah, I think it should flow."

It is important to emphasize that this analysis of communication practices focuses on acting in a specific interactional context rather than on an abstract conception of a particular type of user and what he or she may want generally. It goes further than describing possible personality traits, and instead, draws attention to actions. This analysis suggests that while some users will end interactions with a speech activated system in their car before starting new ones, many participants would prefer to switch between tasks. Unlike when one is interacting with one's smartphone and would perhaps want the phone to "stop listening" altogether at some points—for example, when one places the phone in one's pocket or purse—the preference for switching in the car indicates that perhaps participants do not want their in-car system to stop listening. In other words, participants do not want a change in "degree of access," as Goffman (1972) would describe it (p. 79).

The reasons behind this different preference would need to be explored further. There may be different conceptual boundaries at work here, with system designers employing what we might call an event-based boundary in which a variety of clearly delineated events occur in any given drive, while participants experience a situational boundary, where entering and exiting the car-as-situation signifies the "beginning" and "end." There is also perhaps a parallel to what Schegloff and Sacks (1973) describe as the "continuing state of incipient talk" that occurs between human co-passengers in a car, who can be silent for a period of time without closing an interaction or needing to "begin new segments of conversation with exchanges of greetings" (p. 325). In addition, participants may prefer switching because in one's car, one's primary focus is driving. Consequently, minimal interaction with an in-car speech system, such as that required by switching rather than ending interactions, would likely be more desirable. This preference for fewer steps and, thus, potentially less involved interaction, is also seen in the tendency to use touch versus voice endings, since voice endings require two steps, while touch endings require only one. Participant 9 observes that overall in interacting with the system he feels that "It's easier to touch it, so it's like if I wasn't driving, that would be my choice." However, he is driving and his comment emphasizes that the in-car situation is a complex environment in which one is balancing different activities that require varying degrees of attention and visual engagement at different points, making preferences for interaction perhaps more fluid and dynamic than other contexts, as one switches between driving down the highway or waiting at a stoplight, for example. Murray (2012) observes that every new design medium (for example, the emerging digital medium of an in-car speech enabled system) offers unique affordances, and the design conventions of past media must be changed or adapted to take advantage of these new affordances. Therefore, the conventions for designing past digital media, such as tablets or smartphones, might not be appropriate in designing an in-car system, especially in terms of situational boundaries or demands on visual attention.

8.5 Comparison with findings from mainland China

In our analysis of the data collected in mainland China, we also found that participants engaged in switching from one task to another in Chinese. There were 122 instances of the 25 participants switching between tasks in the Chinese data. We present here a transcript of Participant 4 switching from listening to music to requesting a radio station. Because the participant was previously listening to music, the system initially responds to his touching of the microphone button by asking what song he would like to listen to. However, the system is then able to accomplish the task of switching to a radio station, as the participant requested.

Instance 10

Participant 4 - 23:48

1　P:　((Participant is listening to music. He touches the microphone
2　　　button. Nothing happens. The touch was not recognized.
3　　　Participant touches the microphone button again.))
4　S:　Whose song would you like to listen to? 您想听谁的歌
5　P:　Listen to news 听广播
6　S:　Please repeat 请再说一遍
7　P:　News. FM 103.9. 广播 FM 103.9
8　S:　FM 103.9. Traffic news, is this correct?　103.9FM 交通广播 对吗
9　P:　Correct 对
10　S:　Playing FM 103.9 调台到103.9FM 交通广播

In addition to instances of switching, we also found instances of misalignment when ending tasks in the Chinese data that were similar to the three types of misalignment when ending tasks found in the data from the United States (discussed above). For example, participants forgot to touch the microphone button before attempting to switch tasks. Below is an example of this misalignment from a transcript of Participant 4's interaction with the system.

Instance 11

Participant 4 - 20:46

1　P:　Music 音乐
2　　　((pause))
3　P:　Oh! I have to push the button. 得按一下 ((Participant pushes button.))
4　P:　Music 音乐
5　S:　What type of music would you like to hear? 您想听什么音乐
6　P:　Artist 艺术家
7　S:　Who would you like to listen to? 您想听谁的歌
8　P:　Wu Bai 伍佰
9　S:　Can you repeat that please? 请再说一遍

10 P: Wu Bai 伍佰
11 S: Playing Wu Bai or choose an album 播放 伍佰 或选择一张专辑
12   ((System plays requested artist.))

There were also instances in the data from mainland China of the second type of misalignment identified above in the data from the United States—namely, misalignment that occurred when participants failed to successfully accomplish switching from one task to another. In the transcript below, Participant 1 attempts to switch from listening to music to making a phone call. However, the system is unable to understand him initially and then is slow in processing his request, so the participant ultimately ends the task by pushing the end button and then touches the microphone button and makes his request again.

Instance 12
Participant 1 - 8:22
1   P: ((Participant touches the mic button. There is no ding))
2   P: Phone ca[.... 电话
3   S: [((audible ding))
4   P: Phone call 电话
5   S: Please repeat 请再说一遍
6   (0.2)
7   P: Phone call 电话
8   S: Who would you like to call? 您想给谁打电话
9   P: Contacts 联系人
10 S: ((List of contacts displayed))
11 S: Which contact would you like to call? 拨打哪个联系人电话
12 P: Lu Yen 陆小姐
13   ((Nothing happens on the screen))
14 P: Lu Yen 陆小姐
15 S: Calling Lu Yen on mobile phone 呼叫陆小姐手机
16 P: ((Participant presses the cancel button))
17 P: ((Participant presses the mic button, the system dings))
18 P: Phone call 电话
19 S: Who would you like to call? 您想给谁打电话
20 P: Contacts 联系人
21 S: Which contact would you like to call? 拨打哪个联系人电话
22   ((Participant moves hand toward screen but withdraws))
23 P: Lu Yen 陆小姐
24 S: Calling Lu Yen on mobile phone 呼叫陆小姐手机

In terms of the third type of misalignment identified in the data from the United States, there was one participant from mainland China (Participant 5 from Shanghai), who also attempted to use the Chinese word for "home" to return to the initial screen. This participant used the term "home" a total of seven times during the drive. There were a total of 33 speech requests of "homepage" and 5 speech requests of going "back" in the data from mainland China.

The use of switching in the Chinese data, along with the three types of misalignment described above, suggest that participants recognized similar conceptual boundaries of the car as an ongoing communication situation to those understood by participants in the United States. This finding would indicate that a similar cultural premise is active in both contexts. It would be useful to compare these findings with data from additional cultural contexts to discover if this understanding is common across many cultures or if there are a variety of ways of interpreting the in-car communication situation.

8.6 Conclusion

Researchers suggest that user expectations and preferences when interacting with speech enabled technology differ culturally. System designers make assumptions about the users who will interact with their system and must be careful to take into account different cultural premises of communication. Adopting a cultural discourse theory perspective, we analysed communication practices of task endings in communication between drivers in the northeastern United States and an in-car speech enabled system. Our analysis reveals that while participants ended tasks through touch and voice, they also frequently chose to simply switch tasks rather than explicitly end them. By exploring instances of task endings in which user and system designer expectations appear to have differed, we identified a possible cultural norm *when interacting with a system participant (i.e., the in-car system), if this is done properly, one should be able to switch to a new task without explicitly ending a previous task.* This norm also appears to be active in data from participants from mainland China. We suggest that the expected boundary of the communication situation for users is different from the boundary used in the initial design choices of system designers. It appears that task endings in this context of human-machine interaction in an automobile differ from other contexts of closings between people and between people and other forms of voice activated technology.

Cultural premises about the boundary of the interaction may also inform other premises about sociality and personhood. If the system is conceived as a conglomerate of discrete events, then perhaps one's relationship with that system is also discrete, manifesting and diffusing with the start and end of each event. This in turn may connect to premises about personhood, in which the car is imagined as an instrumental machine to be turned on or off in order to accomplish tasks and nothing more. If, however, the boundaries of the system are situational, and not event-based, then the system is always there, and the relationship continues during and beyond the verbal interaction one has with it, more akin to human-human relation-ships. By extension, premises of personhood may shift with this boundary; the sys-tem is conceived not as an instrument that goes away when it is done completing a task, but as an interactional partner, waiting and listening until you leave. Of course, these premises may not be related in this way in practice—although other research does suggest that people may bring different premises about personhood to their in-car interactions (Molina-Markham, et al., 2016). This is meant only to demonstrate possible connections between cultural conceptions of the boundaries of an interaction and the social relations and models of personhood they implicate. This type of research can lead to improvements or enhancements that are constantly part of the iterative design and redesign process. Thus, it is important to adopt an approach of incorporating findings from research on participant interaction with technology during initial design stages.

References

Berry, M. 2009. The social and cultural realization of diversity: An interview with Donal Carbaugh. Language and Intercultural Communication, 9:230–241.doi:10.1080/1470847090 3203058

Bolden, G. 2008. Reopening Russian conversations: The discourse particle -to and the negotiation of interpersonal accountability in closings. Human Communication Research 34:99–136.

Carbaugh, D. 1988. Talking American: Cultural discourses on DONAHUE. Ablex. (NJ)

Carbaugh, D. 2005. Cultures in conversation. Lawrence Erlbaum. (NJ)

Carbaugh, D. 2007. Cultural discourse analysis: Communication practices and intercultural encoun-ters. Journal of Intercultural Communication Research 36:167–182. doi:10.1080/17475750701737090

Carbaugh, D. 2012. A communication theory of culture. In: (A. Kurylo, ed) Inter/Cultural Communica-tion: Representation and Construction of Culture, Sage. Thousand Oaks, pp. 69-87.

Carbaugh, D., Molina-Markham, E., van Over, B., and U. Winter. 2012. Using communication re-search for cultural variability in human factor design. In: (N. Stanton, eds) Advances in human aspects of road and rail transportation. CRC Press. Boca Raton, (FL), pp. 176–185.

Carbaugh, D., Winter, U., Molina-Markham, E., Van Over, B., Lie, S. and Grost, T. 2016. A Model for Investigating Cultural Dimensions of Communication in the Car. In: Theoretical Issues in Ergo-nomics Science 17(3):304-323.

Carbaugh, D., Winter, U., van Over, B., Molina-Markham, E. and S. Lie. 2013. Cultural analyses of in-car communication. Journal of Applied Communication Research 41(2):195-201.

Coupland, N., and Coupland, J. 2001. Language, Ageing and Ageism. In: (P. Robinson, and H. Giles, eds) The New Handbook of Language and Social Psychology. John Wile. New York, pp. 465–486.

DuFon, M. 2010. The Socialization of Leave-Taking in Indonesian. Pragmatics and Language Learn-ing 12:91–111.

Dogancay, S. 1990. Your eye is sparkling: Formulaic expressions and routines in Turkish. Working Papers in Educational Linguistics 6(2):49–64.

Fitch, K. L. 1990. A ritual for attempting leave-taking in Colombia. Research on Language & Social Interaction 24(4): 209–224.

Geertz, C. 1972. Deep Play: Notes on the Balinese Cockfight. Daedalus 101(1): 1–37.

Goffman, E. 1972. Interaction ritual. Penguin Books.

Goffman, E. 1974. Frame Analysis. Harvard University Press.

Haddington, P. 2010. Turn-taking for turntaking: Mobility, time, and action in the sequential organization of junction negotiations in cars. Research on Language and Social Interaction 43(4):372-400.

Hartford, B. S., and Bardovi-Harlig, K. 1992. Closing the conversation: Evidence from the academic advising session. Discourse Processes 15:93-116.

Hymes, D. 1972. Models for the interaction of language and social life. In: (J. J. Gumperz and D. Hymes, eds) Directions in sociolinguistics: The ethnography of communication. Blackwell. New York, pp.35–71.

Hymes, D. 1974. Foundations in Sociolinguistics: An Ethnographic Approach. University of Pennsylvania Press. Philadelphia (PA)

Knapp, M. L., Hart, R. P., Friedrich, G. W., and Shulman, G.M. 1973. The rhetoric of goodbye: Verbal and nonverbal correlates of human leave-taking. Communication Monographs 40(3): 182–198.

Laurier, E., Brown, B., and Lorimer, H. 2007. Habitable cars: The organisation of collective private transport: Full research report ESRC end of award report, RES-000-23-0758. ESRC. Swindon.

Laurier, E., Lorimer, H., Brown, B., Jones, O., Juhlin, O., Noble, A., Perry, M., Pica, D., Sormani, P., Strebel, I., Swan, L., Taylor A. S., Watts, L., and Weilenmann, A. 2008. Driving and passengering: Notes on the ordinary organisation of car travel. Mobilities 3(1):1-23.

McKay, E. N. 2013. UI is communication: How to design intuitive, user centered interfaces by focusing on effective communication. Elsevier. Waltham (MA)

Molina-Markham, E. 2014. Finding the 'sense of the meeting': Decision making through silence among Quakers. Western Journal of Communication 78(2):155–174.

Molina-Markham, E., van Over, B., Lie, S. and D. Carbaugh. 2016. "You can do it baby": Cultural norms of directive sequences with an in-car speech system. Communication Quarterly 64(3):324-347.

Murray, J.H. 2012. Inventing the medium: Principles of interaction design as a cultural practice. MIT Press. Cambridge.

Nass, C. and Brave, S. 2005. Wired for speech. How voice activates and advances the human-computer relationship, MIT Press.

Nass, C. I., and Yen, C. 2010. The man who lied to his laptop: What machines teach us about human relationships. Current. New York (NY)

Omar, A. S. 1993. Closing Kiswahili conversations: The performance of native and non-native speakers. Pragmatics and Language Learning 4:104–125.

Philipsen, G. 2002. Cultural communication. In: (W. Gudykunst and B. Mody, eds) Handbook of international and intercultural communication. Sage. London and New Delhi, pp. 51-67.

Sacks, H., Schegloff, E. A., Jefferson, G. 1974. A simplest systematics for the organization of turn-taking for conversation. Language 50:696-735.

Schegloff, E. A., and Sacks. H. 1973. Opening up closings. Semiotica 8(4):289–327.

Scollo, M. 2011. Cultural approaches to discourse analysis: A theoretical and methodological conversation with special focus on Donal Carbaugh's Cultural Discourse Theory. Journal of Multicultural Discourses, 6: 1–32. doi:10.1080/17447143.2010.536550

Scollon, R., Scollon, S., 1981. Narrative, literacy, and face, in interethnic communication. Ablex. Norwood (NJ)

Takami, T. 2002. A Study on Closing Sections of Japanese Telephone Conversations. Working Papers in Educational Linguistics 18(1):67–85.

Townsend, R. 2009. Town meeting as a communication event: Democracy's act sequence. Research on Language and Social Interaction 42(1):68–89.

Turkle, S. 2011. Alone together: Why we expect more from technology and less from each other. Basic Books. New York (NY)

Van Over, B. 2014. Tracing the decay of a communication event: The case of the daily show's 'seat of heat.' Text & Talk 34(2):187–208.

Winter, U., Tsimhoni, O., and T. Grost.2011. Identifying cultural aspects in use of in-vehicle speech applications. Paper presented at the Afeka AVIOS Speech Processing conference, Tel Aviv, Israel.

Winter, U., Shmueli, Y., and T. Grost. 2013. Interaction styles in use of automotive interfaces. In: Proceedings of the Afeka AVIOS 2013 Speech Processing Conference, Tel Aviv, Israel.

9 Communication and Cultures in Cars: Reflections and Looking Forward

For some time now those who have been concerned about human-computer interaction have struggled with conceptualizing and investigating cultural variation. An ultimate hope in this has been to create user interfaces that honor cultural practices but which also cross cultural communities in highly desirable ways. Attempts to research the role of culture in design and usability have often involved the use of abstract, macro-level dimensions such as those provided by Hofstede (1991), Hall (1989), or Ting-Toomey (1998). These dimensions are attractive because of their purported universality, and the ease with which one can begin to design for a given population once one knows if that group is, for example, a "high" or "low" on some variable such as context, long term orientation, and a variety of other such attributes (for a discussion of this see Carbaugh, 2015). However, such approaches have recently been criticized for their tendency to essentialize diverse groups along these dimensions while also stereotyping populations around national boundaries; the approaches have also failed to effectively guide the development of culturally satisfying interfaces (Winschiers & Fendler, 2007; Sun, 2009). In other words, it is unclear how these broad dimensions can consistently inform the design of user interfaces that require the production of particular communicative actions and sequences in which real people will interact.

As we reflect upon our studies included above, we find it valuable to review the kind of problems our studies have raised and addressed. In chapter three, we explore what a technology does when it is used in ways for which it was not designed. How does a system respond when non-task features are introduced to it, especially when at times users like it to be "off task"? In chapter four, we find there are times when participants want maximum efficiency when completing a task, but there are other times when users want a more interactive conversational partner. How do we understand that and how do we design our communication with that knowledge? In chapter five, we explore what happens when basic turns taken in interaction counter the expectations and preferences of users. What does one do about that? In chapter six, we find considerable overlap in utterances between users and the car's system at particular times, and less so at others. How is this understood and what should be done about it? In chapter seven, we find apologies and wonder why they are being used and what they are designed to achieve. How can the interactional system be better designed based upon this finding? In chapter eight, we explore how participants switch among tasks, including ending of specific communication events.

The specific problems we raise throughout, are themselves findings of our research. In other words, we did not begin by focusing upon each, but discovered each as a result of our studies. Without this sort of stance toward qualitative inquiry, we would not have found the dynamics reported. In the process we bring to the fore,

https://doi.org/10.1515/9783110519006-009

specific communication acts such as off-task talk and apologies, act sequences such as taking and breaking turns as well as overlapping talk, and communication styles such as optimal efficiency or maximal interactivity. We explored what each means to participants as well as their preferences when each is active. This holds in view, then, cultural features in the interactions as the nature and meaning of these communication events are brought to the fore.

The approach we have taken to address these problems is different from those that focus on abstract dimensions and aggregate scores of populations. We are not assessing individual-collective dynamics but cultural features in interactions, participants' sense of the communication and its meanings. Our approach hones in on actual communication practices in cars while exploring cultural features of those practices – in the United States and China - including their design. This stance to inquiry is based in the Ethnography of Communication (Hymes, 1962, 1972; Philipsen, 1992), and one of its more recent developments, Cultural Discourse Analysis (CuDA). The framework is a general theory yet yields, when used, a bottom-up understanding of usability rather than a top-down view based in abstract dimensions. We believe the utility of such an approach lies in its ability to identify cultural models of design with these being based upon local conceptions of interactions, persons, relations, emotion, and place, as well as the cultural preferences for channels and instruments, all of which are required for the understanding and development of truly localized practices (Carbaugh, 2007, 2008; Scollo and Milburn 2019).

In addition to outlining how such a theory and method can be used effectively to address issues of design and usability in HCI, we have presented a research procedure for the investigation of human-computer interactions in an automobile. As we summarized in chapter one, in our initial work, GM investigators discovered and discussed how cultural dimensions of activities were not only possible but perhaps even pervasive throughout various automotive speech applications (Tsinhomi, Winter, and Grost, 2009). These researchers noted how cultural features were a key part of the utterances people produce in a car, in the vocabulary they used, the variety of sounds they made as well as the variation in the phonemes produced. As we have seen throughout this volume, the importance of these features became apparent in several communication practices including the flow of conversation, response times, how prompts were formulated, and how errors were handled. The role of cultural variability in automotive communication then, is pervasive and powerful, yet understudied if at all. Such work, in our assessment, and in our design, is necessary.

In this final chapter, we will briefly review the field research cycle we introduced in chapter two. We review the type of research design we have used in our studies, followed by a set of considerations for establishing field sites, including both the means of collecting data and the procedures for its analysis. We will construct our design here as a reflection on what we have done for the studies gathered here as well as what others can do if desiring to conduct similar inquiries. Our view then is both reflective and forward looking.

9.1 Field Research Cycle

The framework for field-based, cultural research into communication that we used here has involved a sequential design. To begin, we will summarize this briefly in four phases, from the activities we did prior to entering the field, to activities completed after leaving the field. This overview will make explicit the general foundation for a discussion of the specific applications of the framework we used to research in-car communication including the various means of collecting and analyzing naturalistic data in this context.

9.1.1 Pre-fieldwork Activity

This phase of the project involved several activities, which are preliminary to doing the fieldwork itself. Before entering, it is crucial to be as knowledgeable about a field site as is possible. If there is a literature available about the site, or various participants' practices, researchers should consult it in order to become more knowledgeable about the history of the area, its people, their occupations, customs, and so on. We caution that the research may find more and/or other than these, reported earlier, but nonetheless it is valuable to explore this literature prior to entering the field.

A second set of activities involves detailed planning about the fieldwork itself. If there is a special focus envisioned, then a preliminary conceptual map of that focus should be formulated. For example, if one were most interested in users' tactile manipulation of a navigation system, then what that is and how it works should be carefully thought through. If one were interested in interruptions, then that practice should be carefully conceptualized (as above in chapter 6) and so on. Focusing on specific concerns in this way establishes a theoretical position from which to observe communication in a particular context. This equips one for study and reflection while in the field, enables a systematic approach to one's observations, as well as provides ways of designing interviews, analyses, and eventually the practice itself.

9.1.2 Fieldwork Activity

This phase involves periods of observation of communication in its actual cultural context(s). Observations such as these inevitably lead to questions which researchers can ask participants during interview sessions as we note throughout the above chapters. The accumulation of a descriptive record about observational and interview events creates a corpus of data which then is subjected to analysis. The analyses involve the distinct modes of investigation that we will discuss in the next sections.

9.1.3 Post-fieldwork Activity (Stage One): Descriptive and Interpretive Analyses

The activities conducted after leaving the field involve continuing phases of analysis, which involve: 1) descriptive analyses about how particular communicative activities are done in a specific cultural context and 2) interpretive analyses about what the meanings of those activities are for the people who engage in them. These analyses lead, often, to additional questions about dynamics one observed, or heard about while in the field. These analyses of the data can - and typically do - lead back to the field for more detailed observations.

9.1.4 Post-fieldwork Activity (Stage Two): Comparative Analyses and Critical Assessments

This final stage of a field project is crucial as a sharper view of the cultural dimensions of communication get better understood through comparative analysis. How is the practice such as interruptions to the car's system, similar to, yet different from, other practices elsewhere? Also, through critical study with participants, researchers can contribute ideas about what users take to be better design of these practices, products, policies, or other human creations (Carbaugh, 2008).

Note, and we emphasize, that the stages and analyses we present here are sequential in their design but cyclical in their possibilities. Some period of fieldwork can result in revising one's earlier conceptualizations; post-fieldwork activities can lead one back to the field focused on other observations, with different questions, and so on.

9.2 Some Considerations of the Field Site

The framework described above has been used in a variety of contexts to study culturally-informed communication (e.g., Carbaugh, 1988, 1996, 2005; Saito, 2009; Boromisza-Habashi, 2011; see the recent review in Scollo and Milburn, 2019). Here we describe a model for applying this framework to an examination of communication in automobiles. This description is the basis for future research that has been undertaken and could be further explored at various locations around the world.

9.2.1 The Field Sites

Our field sites in the United States and China were selected in order to enable comparison of driving practices in different cultural contexts. These could include not only national variation but have also included rural and urban driving in various metropolitan areas. As we have discussed, cultural comparison is an essential phase of investigation in Cultural Discourse Analysis and the selection of a variety of field sites facilitates this analytic process.

9.2.2 Selection of Participants

Several criteria guided and can guide the selection of participants for each field site of the study. A pre-questionnaire could solicit basic demographic information about each potential participant, about their willingness to have their data downloaded into the automobile computer system, and other information concerning their driving experiences and habits. This preliminary information was and could be further used to ascertain the suitability of a participant for the study. Additionally, and most importantly, most participants of the study should be a native speaker of the language in which the speech application has been designed. This is important as a control on the speech recognition and dialog flow capabilities of the system being tested. Additional criteria to be met could include some balance between male and female drivers, tech savvy and not, young and old, urban and rural, and if possible, including early adopters of technology. We note language proficiency could also be added as an additional variable if desirable.

9.2.3 General Schedule at each Field Site

In this section we provide a detailed discussion of the steps of the research cycle described above as they could be undertaken in studying communication in automobiles.

Following our studies above, we envision a potential pre-fieldwork period of 10-14 days which would involve the research team in contacting possible participants and making local arrangements. An intensive fieldwork period can be designed to occur over a 10 day period which could be set as follows:

Day 1: Arrival, settle, plan and meet with local contacts, establish possible routines, conduct initial observations of settings.

Day 2: Conduct a preliminary pilot study of 1-2 participants using the *a priori* framework; do some very tentative descriptive and interpretive analyses, make any revisions that may be needed in the research design.

Day 3: Observational work in the car with 2 participants (recorded); interview afterward; conduct preliminary analyses of initial data; review and revise the framework as needed.

Days 4-8: Conduct driving, interview, and debriefing sessions with 2 participants each day including observations in the car; preliminary analyses of data; eventually integrating data and analyses; create an initial descriptive account; revise framework as needed

Day 9: Final data gathering (as needed); final in-field analyses (translations, interpretations).

Day 10: Depart.

A post fieldwork period of 10-20 days would involve the development of the descriptive reports about the communication including analyses of specific sequences and norms. This would also include the development of a very preliminary interpretive report of the cultural terms, sequences, meanings, and aesthetics of the car as a communication situation; assessments of theory, methodology, and findings.

9.3 Means of Data Collection

Drawing on the framework described above, multiple schematic ways of collecting data become possible including an observational scheme, an interview guide, and technical logs.

9.3.1 Gathering Data from Study Participants

Upon first meeting the research team, selected participants were introduced to the study, and given a pre-questionnaire to assess their suitability for the study. If meeting the inclusion criteria, the participant was asked to sign a consent form, which made explicit conditions for participating in the project. Each was then introduced to the capabilities of the car's speech recognition and dialog system, would engage in an off-road (i.e., parking lot) test drive, would engage in a driving session, and finally would conclude with an interview session. The overall session with each participant lasted around 120-150 minutes.

9.3.2 Field Researchers:

In the context of this particular application, the primary task of researchers would be to monitor participants' driving and to formulate questions about the driving to be asked in the subsequent interview session. During the "drive-along," researchers would be advised not to interrupt the driver while driving, but they could respond to questions the driver asked. The researchers also were involved in sharing observations about the drive-along during subsequent interviews and during a debriefing session among researchers that was held after each driving-interview session. Driving sessions involved one or two or at times even three researchers in the automobile with the driver.

9.3.3 Observational Scheme

The observational scheme equips the researcher with a specific, structured way of watching and listening while in the car with study participants. One such scheme based in the ethnography of communication (Hymes, 1980) involves selective attentiveness to the following components:

1. **Setting**: In what physical environment is the communication taking place?
2. **Participants**: Who is involved in the communication practice?
3. **Ends**: This component has two parts: What are the participant's goals of the practice (e.g., to send an email)? What are the outcomes of the practice (e.g., the email was sent, or the effort to do so was unsuccessful, or the user got irritated at the car)?
4. **Act/Sequence**: What specific communication acts got done, and in what sequence?
5. **Key**: What was the emotional pitch, or tone of the communication (e.g., perfunctory, serious, frustrated)?
6. **Instrumentalities**: What multiple mode(s) or cues were used in this communication (e.g., voice, gesture, pressing a button, words)?
7. **Norms** (see below): What were the norms or participant preferences - stated and/or implied - for this interaction?
8. **Genre**: Is there a generic form to this communication practice which participants use, and if so, what is it?

These eight components provide a basic investigative tool for analysts to systematically describe communication in and about the car. As is illustrated in the above chapters, a sub-set of the concepts could be more useful in some cases than in others.

Based upon prior works, we find a recommended way of using the observational scheme involves creating a chart on a piece of paper (or recording device), to be filled in by the researcher. This scheme could be used by the researcher with each component across the top of the paper and a time-line down the left margin, so the researcher could record preliminary observations about particular components at specific times. For example, during the 21st minute (recorded on the left side) the driver attempted to change radio stations (the A, I-through touch, K-frustrated) but failed (the End). These observations could then be used to formulate questions during the later interview session. Each can also suggest focal data for preliminary analyses.

9.3.4 The Interview Session

After the driving session, the researchers asked the driver a series of questions. An interview guide was provided based upon the framework described above. The wording of the questions was shortened and modified to fit the particular field participants and sites of concern.

The first section addressed the driving session itself, and anything which had been noticed during it and warranted probing. A second set focused on the participant's (the driver's) background using dialog systems. In addition to giving information about the participant, these questions established a wider context for the particular communication situation of the driving session. Next, the questions focused on the sequencing of the driving session –for example preferences for how a dialog should be initiated, the degree and form of feedback that the system should provide while a particular exchange is taking place, the repairing of communication errors, and the ending of the interaction. The final questions were designed to explore various modes of cueing that the driver engaged with the system; these modes may or may not have come up during the discussion of sequencing and these questions thereby encouraged further exploration of those that may not (yet) have been discussed.

The questions in each section were designed to prompt the participant to produce talk about dialog systems that could be analyzed for cultural propositions and premises that the participant used to interpret and produce their interactions. Questions that more explicitly explored cultural premises the participant may hold about the ideal driver (or system) were posed, including relations with the interface as well as others in the car, feelings about the car/interface, and in-car communication generally. The interview session closed with a catch-all question that would allow the participant to share anything about their experience that was not explicitly requested in prior questions.

9.3.5 The Debriefing

In the end, the researchers met together to discuss the overall session. The purposes of the debriefing session were to reflect upon the specifics of the driving session, the interview session, particular observations made during each, to identify useful focal concerns for further attention, perhaps to modify the methodology in subsequent sessions, and so on.

9.3.6 Technology

For any particular study, the car's human-machine interface could be designed to explore a system in general, or specific features of any one system. It is desirable, as we did, to add logging capabilities to enrich the corpus of data for subsequent analyses. For exploring spoken communication with the described methodology, the HMI needs to contain a speech application as part of a multimodal interface, which supports the use of natural language utterances and a pre-defined variety of typical in-vehicle tasks. This allows examining cultural variation of any verbal, as well as non-verbal acts as part of the communication.

9.4 Procedures for the Analysis of Data

After collecting the data via the above methods, there are specific procedures that are employed for their analysis. The ones we use are summarized here in four phases.

9.4.1 Phase One: Descriptive Analysis

A descriptive phase of study would respond to the question, how is that particular communication practice getting done here? What are exact instances of the phenomenon of concern to participants in this real time and place? Care is taken to record multiple instances of the phenomenon of interest such as selecting music, making a call, errors and their (attempted) correction, interrupting, or direction giving. Detailed descriptions are created of each instance – based upon the audio and video recordings. A group of instances would comprise a focused corpus for study. Note that the analytic tasks here following recording of data and then would involve the following: 1. noticing an instance of a phenomenon; 2. making a collection of multiple instances of that phenomenon; 3. transcribing the instances so that a descriptive record of the pattern is created.

After creating a descriptive record, researchers are able to search across multiple instances for linguistic and non-linguistic qualities which recur in a patterned way. The descriptive analysis would lead to claims such as: This is how this practice such as direction giving, or attempted error correction, or volume control, or implementation, or speech prompting is actually getting done in this corpus of data.

A key part of the descriptive analysis is the recording of observational data onto electronic devices (such as video and audio recorders) and the subsequent transcription of those data. The latter are done using specific transcription conventions and orthographic techniques. These include the ways nonverbal positions and movements are active during the activity being studied. Also synchronized with these would be the verbal data logged into the system at the exact time of the instance. The careful documentation of these data and their inscription would provide the toe-hold of the field study in actual communication events or sequences.

9.4.2 Phase Two: Interpretive Analysis

An interpretive phase of study would respond to the question, what does this specific communication practice (or finding from the descriptive analyses above) mean to participants? What cultural significance and value (or lack thereof) does this have for them? While a descriptive analysis gives a field investigator evidence of patterns of practices people create together; an interpretive analysis tells us the meanings of those practices. An interpretive analysis would follow exacting procedures which take different trajectories, but in all cases, this analysis would begin:

1. take one pattern of practice that has been documented and analyzed through descriptive study;
2. examine the corpus of instances of that practice;
3. if available select participants' terms which are active in the corpus and use them to formulate a statement about that practice (the participants' terms are called cultural terms and should be placed in quotation marks; the statement formulated through *cultural terms* is called a *cultural proposition*).
4. if salient to participants, trace the meanings of this proposition about the action getting done, the feeling of it, identity issues, social relations, emotions;
5. interpretation of norms: What is being presumed about good or proper practice in this phenomenon? This involves a four-part analysis, through which specific communication norms can be explicated - and later compared - in a prototypical formula which involves these four parts:
   a. Context: When done (the setting or context of the car, with participants P, or participant relations R),
   b. Condition of Identity (relation to conduct): if one wants to (e.g., be a good driver, get directions to a place),
   c. Force: one (must, preferably should, permissibly could, must not),
   d. Conduct: do action X.

Communication norms are analyzed further by positing explicit or implicit imperatives, along dimensions of intensity (how strong participants feel about the norm) and crystallization (the amount of agreement about the norm).

6. additional trajectories of interpretive analyses build on the above, and can be elaborated by examining cultural metaphors, semantic dimensions or cultural premises.

These analytic procedures would provide bases for claims about the meaningfulness of practices to participants.

9.4.3 Phase Three: Comparative Analysis

A comparative phase of study would respond to the questions: to what degree is this practice the same and to what degree is it different from other practices? If one were to study direction-giving in Atlanta and in Shanghai, or speech prompting of entertainment, we would expect some similarities and some differences in this practice in the two locations. Through such study we could get a better idea about what indeed is culturally distinctive in one set of practices, as opposed to those in another; we also would get a better sense of what is similar across such practices. Through comparative study, and based upon descriptive and interpretive analyses, one would be well-placed for such assessments.

9.4.4 Phase Four: Critical Analysis

A critical assessment would begin by asking what works well and what does not? After carefully investigating practices through the above modes, researchers could better reflect upon what is working well and what is not. Critical assessment and subsequent considerations could lead to better designed capacities, and, in this case, more usable applications.

9.5 Summary

The framework presented here has provided the investigative stance for our studies; in other words, this is our model for ethnographic investigation of issues of design and usability in human-computer interaction in automobiles. Applying this framework to a particular context, we have outlined cyclical phases of fieldwork activity, schematic means for collecting data, and procedures for the analyses of these data. The success of a speech application depends greatly on its situated use in a particular cultural context. Our framework seeks to uncover the distinctive cultural models of communication that are active in such contexts, and various communities, to ascertain means of adapting human-machine interfaces to local contexts of use.

The framework for naturalistic field studies is designed to be seamlessly integrated into a user-centered design approach (Dray & Siegel, 2009; Norman & Draper, 1986). It should clearly be part of the research phase before any design, implementation and evaluation cycle. Thus, designers have an overview on cultural norms and preferences for all necessary interface design dimensions, such as cooperative principles for communication, control handling, dialog patterns and sequences, turn taking and grounding conventions, information distribution, choice of words and phrases in dependency of typical automotive tasks and domains, and conflict resolution. We offer it as such and hope to contribute further on its bases in our future studies.

References

Berry, M. 2009. The social and cultural realization of diversity: An interview with Donal Carbaugh. Language and Intercultural Communication, 9:230–241.doi:10.1080/1470847090 3203058

Boromisza-Habashi, D. 2011. Dismantling the antiracist "hate speech" agenda in Hungary: An ethno-rhetorical analysis. Text &Talk 31:1-19.

Carbaugh, D. 1988. Talking American: Cultural discourses on DONAHUE. Ablex (NJ)

Carbaugh, D. 1996. Situating Selves. State University of New York Press. Albany (NY)

Carbaugh, D. 2005. Cultures in conversation. Lawrence Erlbaum (NJ)

Carbaugh, D. 2007. Cultural discourse analysis: Communication practices and intercultural encounters. Journal of Intercultural Communication Research 36:167–182. doi:10.1080/ 17475750701737090

Carbaugh, D. 2008. Putting policy in its place through cultural discourse analysis. In: (E. Peterson, ed) Communication and Public Policy Proceedings of the 2008 International Colloquium of Communication. Digital Library and Archives, University Libraries, Virginia Tech, pp. 55-64.

Carbaugh, D. 2015. Intercultural communication as a situated, culturally complex, interactional achievement. Russian Journal of Linguistics 19:33-42.

Carbaugh, D., Molina-Markham, E., van Over, B., and U. Winter. 2012. Using communication research for cultural variability in human factor design. In: (N. Stanton, eds) Advances in human aspects of road and rail transportation. CRC Press. Boca Raton, (FL), pp. 176–185.

Carbaugh, D., Nuciforo, E. V., Molina-Markham, E. and van Over, B. 2011. Discursive reflexivity in the ethnography of communication: Cultural Discourse Analysis. Cultural Studies <-> Critical Methodologies 11(2):153-164.

Dray, S. and D. Siegel. 2007. Understanding users in context: An in-depth introduction to fieldwork for user centered design. In: (C. Baranauskas, P. Palanque, J. Abascal, S.D.J. Barbosa, eds) Human-Computer Interaction – INTERACT. Lecture Notes in Computer Science. Springer. Berlin Heidelberg, 4663: pp. 712-713

Hall, E. T. 1989. Beyond culture. Doubleday. New York (NY)

Hofstede, G. 1991. Cultures and organizations: Software of the mind. McGraw-Hill. London.

Hymes, D. 1962. The ethnography of speaking. In: (T. Gladwin, and W. Sturtevant, eds) Anthropology and human behavior. Anthropological Society of Washington. Washington (DC)

Hymes, D. 1972. Models for the interaction of language and social life. In: (J. J. Gumperz and D. Hymes, eds) Directions in sociolinguistics: The ethnography of communication. Blackwell. New York, pp.35–71.

Norman, D. and S. Draper. 1986. User centered system design; New perspectives on Human-Computer Interaction. Lawrence Erlbaum. Mahwah (NJ)

Philipsen, G. 1992. Speaking Culturally. State University of New York Press. Albany (NY)

Philipsen, G., and L. M. Coutu. 2005. The ethnography of speaking. In: (R. Sanders and K. L. Fitch, eds) Handbook of research on language and social interaction. Lawrence Erlbaum Associates. Mahwah, pp. 355–379.

Saito, M. 2009. Silencing Identity through Communication: Situated Enactments of Sexual Identity and Emotion in Japan. VDM Publishers. Germany.

Scollo, M. 2011. Cultural approaches to discourse analysis: A theoretical and methodological conversation with special focus on Donal Carbaugh's Cultural Discourse Theory. Journal of Multicultural Discourses, 6: 1–32. doi:10.1080/17447143.2010.536550

Scollo, M. and Milburn, T. 2019. Introduction. Engaging and Transforming the World through Cultural Discourse Analysis. Fairleigh Dickinson University Press.

Sun, H. 2009. Designing for a dialogic view of interpretation in cross-cultural IT design. Lecture Notes in Computer Science 5623:108-116.

Ting-Toomey, S. 1989. Intercultural conflict styles. A face-negotiation theory. In: (Y. Y. Kim and W. B. Gudykunst, eds) Theories in intercultural communication. Sage. Newbury Park.

Tsimhoni, O., Winter, U., and Grost, T. 2009. Cultural considerations for the design of automotive speech applications. In: Proceedings of the 17th World Congress on Ergonomics IEA 2009, Beijing, China.

Winschiers, H. and Fendler, J. 2007. Assumptions considered harmful. Human Computer Interaction 10:452-461.

Winter, U., Tsimhoni, O., and T. Grost.2011. Identifying cultural aspects in use of in-vehicle speech applications. Paper presented at the Afeka AVIOS Speech Processing conference, Tel Aviv, Israel.

10 Appendix A

Fig. 1: Home Screen – English

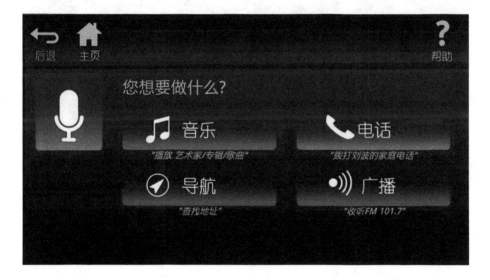

Fig. 2: Home Screen – Chinese

https://doi.org/10.1515/9783110519006-010

Fig. 3: Contacts Screen – English

Fig. 4: Contacts Screen – Chinese

Fig. 5: Music Screen – English

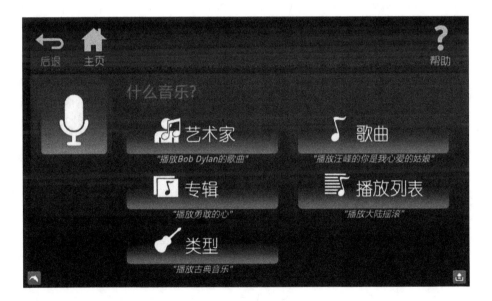

Fig. 6: Music Screen – Chinese

Fig. 7: Radio Screen – English

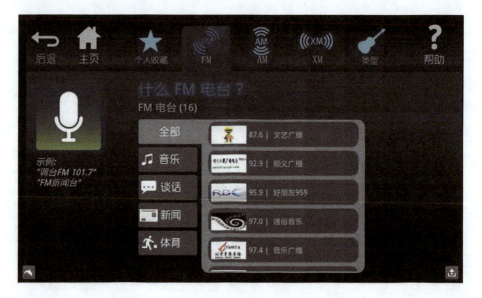

Fig. 8: Radio Screen - Chinese

www.ingramcontent.com/pod-product-compliance
Lightning Source LLC
LaVergne TN
LVHW062318060326
832902LV00013B/2285